A Chanticleer Press Edition

The Secret Life of Animals

Pioneering Discoveries in
Animal Behavior

Lorus and Margery Milne
Franklin Russell

E.P. Dutton & Co., Inc.
New York

Published in the United States by
E. P. Dutton & Co., Inc.
New York

Prepared and produced by Chanticleer Press, Inc.
Staff for this book:
Publisher: Paul Steiner
Editor-in-Chief: Milton Rugoff
Managing Editor: Gudrun Buettner
Associate Editor: Jeffrey Simpson
Picture Editors: Ann Guilfoyle, Eve Kloepper
Production: Helga Lose, Judy Shugerman
Design: Massimo Vignelli

Manufactured by Dai Nippon Printing Co., Ltd.,
Tokyo, Japan

Library of Congress Catalog Number 75-10073
ISBN 0-525-19932-2

Preceding portfolio: In the afternoon sun an African lion
mates with one of the females in his pride. *(Photographs
by Fran Allan/Animals Animals)*

Contents

Introduction

Only about twenty years ago the lion was thought to be brave, fierce, indomitable, and powerful enough to subject most other animals to his will. When the lion thrived, it was believed, all other animals suffered. The zebra, the wildebeest, the gazelle, and the impala were his natural victims, and he controlled their numbers. Today, however, the so-called king of beasts is known to be a clumsy hunter who has little effect on the populations of his prey. As a nomad, he scavenges much of his food, driving away vultures and solitary hyenas. As a pride animal, he hunts hardly at all but appropriates the kills of his lionesses. He can be driven from his meal by packs of hyenas.

Twenty years ago the gorilla was thought to be almost equally fearsome, but today he is known to be a timid and harmless vegetarian. Until very recently the hyena was believed to be a scavenger, a skulking coward, the object of almost universal human revulsion from the Bay of Bengal to the southern tip of Africa. Today he stands revealed as a bold and ingenious killer.

Throughout the world, and throughout history, animals have been regarded anthropocentrically—that is, as inferior life forms which perforce must circle helplessly around the all-powerful human being. Since man is dominant on earth, he also must be its most superior life form, as evidenced by prodigious deeds in art, literature, and science. It is sobering today to look back over these last twenty years and see the great revolution that has taken place in man's attitude toward animals. Instead of believing in his anthropocentric position in the universe, he is now increasingly concerned that he may be a helpless victim, not much more than a pawn in a cosmic game he does not understand. He suspects that he really has no control over his life and certainly only marginal knowledge of its intricacies. This change in attitude has come about subtly. While rockets have explored other planets, while hearts and kidneys have been transplanted, while new advances in agriculture, sociology, physics, and medicine have occurred, students of animal behavior, the ethologists, have painted an entirely different picture of man's occupancy of his earth. In this picture, he is just another animal. Moreover, he is an animal who subjugated his world before he was capable of controlling, or even understanding, himself.

We have no information that any of the ancient civilizations of Babylon, Assyria, Sumer, or Egypt had much concern about the modes of life or the secrets of the animals around them. Rather, animals were chattels at the disposition of man's will. Many large African mammals were brought by the shipload from the fabled land of Punt to decorate the irrigated gardens of Egyptian pharoahs. Lions were captured by blood-thirsty Assyrians and pitted against heavily armed men. Greek bullfighters fought bulls to the death. Lions were used to kill Christians in the Roman arena and monkeys became pets of Elizabethans. Animals were tamed and exploited by all civilizations and there is no hint anywhere that men asked the why, the wherefore, or the how of their behavior.

Aristotle is the first man we know to have inquired into the secret lives of animals. An insatiable watcher of all things, he gathered specimens on the seashore, observed the migrations and nesting habits of birds, and studied the behavior and movements of all the animals he saw. He dissected and measured and kept asking himself why certain animals lived only in certain places, and why their behavior followed predictable patterns. His *Historia Animalium* is not remarkable in comparison with his other writings. But, incredibly, it stimulated no similar inquiry into animal behavior for the next two thousand years. After Aristotle's pragmatic investigations came centuries of romancers and fabulists who looked upon animals as extensions of man's imagination and as symbols of his grosser faults.

In the seventeen hundred years between Aristotle and René Descartes, the French philosopher, man advanced only to the thesis that animals were automata, incapable of thought or feeling, and without relevance to the human condition. Even today there remains one fossilized example of this attitude. In South Africa there is a legal ruling to the effect that cold-blooded animals cannot feel pain, thus denying them the protective jurisdiction of the Society for the Prevention of Cruelty to Animals.

Animals, presumed to be lacking in any qualities even faintly connected with human life, long remained convenient symbols for other things. The fabulous unicorn recurs throughout medieval history as a symbol of strength, and its horn, which supposedly decorated cups, was believed to nullify the effects of poison. The ancient mind was preoccupied by monsters and devils in animal form, by strange and fearsome creatures perpetually landlocked in deep lakes, and by whales big enough to eat ships.

The first man to challenge the dogmatic view of animals as living machines was Charles Darwin, in his *Expression of the Emotions of Man and Animals,* published in 1872. It was a brilliant and heretical exposition, which related man and animal in a way that only a mind capable of conceiving the theory of evolution could write. He created the conditions that enabled man finally to accept his own dumb animal past. He took the first step into the labyrinth that we are now struggling to master. The earliest attempts to probe the mysteries of animal behavior were restricted to the more primitive forms of life. In the eighteenth century Luigi Galvani, an Italian physiologist, used electricity to cause the muscles in a frog's freshly severed leg to twitch. The experiment was, however, less concerned with animal behavior than with relating electrical energy to life. One hundred years later, an obscure French schoolteacher, Jean Henri Fabre, took the study of animal behavior to its first plateau, from which all later work would be measured.

Before Fabre, most observations of animals had been on the Aristotelian model: the watching of individual acts, the recording of the disappearance and reappearance of migratory animals at certain times of the year. It was raw data collected at random. Fabre integrated the lives of the insects he was watching into systems that made sense in human terms, while preserving their independence from anthropomorphic judgments. Between 1852 and 1907 he was almost always in the field while other students of animal behavior were cutting and drugging and electrifying their subjects in laboratories.

While Fabre was prowling the French countryside, and was being dubbed insane by friends and neighbors, Sir John W. Lubbock was working in England on the mystery of the "sense of color" possessed by many of the lower animals. He proved that honeybees prefer the color blue, that they opt for bright sunlight, and that they work best when the temperature is around 75 degrees.

Lubbock's work suggested subtle elements in the organization of even simple forms of life. At the same time, Jacques Loeb, a German physiologist working in America, was trying to measure the reactions of animals to light, to gravity, to wind, and to electricity. So predictable were the responses of both animals and plants that he came to regard them as tiny machines incapable of reacting independently and individually to the world around them. He published his classic book, *Forced Movements, Tropisms and Animal Conduct,* in 1916.

As the experimenters pushed ahead, however, nagging doubts grew that the laboratory environment might dictate artificial behavior. Loeb, showing typical ingenuity, varnished the eye of a fly. The untreated side of the fly's body stiffened and the entire body tilted when the fly tried to walk. It turned in circles and collided with objects on its blind side. But after being left in a box overnight, the fly proved to be perfectly capable of walking on its six legs without tilting. In just one night it had adjusted to the partial blindness Loeb had inflicted.

All students of animal behavior were confronted by one major problem: namely that behavior is a response and something must trigger it. The trick was to either locate the trigger, and define it, or to know when or how the trigger would occur. For the field worker, this was a big enough problem but the laboratory worker could never be sure whether his artificial trigger was not influenced by the laboratory conditions or that the laboratory animal was not already under the artificial influence of its confinement.

An American naturalist, Herbert S. Jennings, preferred to work with animals so simple that they were insensitive to the artificial environment of the laboratory. He used protozoans, single-celled organisms, and found them capable of discrimination and adjustment to the outside world, just as Loeb's fly had been. Jennings believed that this flexibility of response was part of the inherited repertoire of all animals.

In his classic *The Behavior of Lower Organisms,* he reported that despite the simple forms of the animals, almost any change in the environment produced a response. Some protozoans swam away from strong sunlight, but would swim toward the light if it were weakened. They "remembered" the position of the light and would swim toward the place where light had shone after it had been cut off. His work illuminated one fact of behavior: The response of the animal was often its attempt to adapt to a changed condition. And although the principle appears simple, it was given force by its manifestation in such primitive creatures.

Around this time, German scientists were anesthetizing dogs and shocking them electrically in an effort to plot their nervous connections. Germans then found that electrochemical messages passed from sensory centers through the spinal cords and brain, then to muscles and glands, affecting their responses. Sir Charles Sherrington mapped many of the functional tracts of the spinal cord and the brain in the early 1900's. He discovered sense organs in skeletal muscles and isolated the synapse, that physiochemical reaction which occurs at the contact point of nerve cells. Another British scientist, Edgar D. Adrian, who won the 1932 Nobel prize with Sherrington in physiology and medicine, worked on measuring nerve impulses and electrical rhythms in the brain. He was able to amplify and record nerve impulses from sense organs. These studies were necessary for man to discover how little he actually knew. But there was still barely any recognition of the great depths of behavioral mystery in the lives of animals.

The great Russian physiologist, Ivan Pavlov, whose work was based almost entirely on dogs, is remembered today as the discoverer of the "conditioned reflex" (or "conditioned response"). But his work actually went far deeper than most scientists then suspected. By a complex system of punishments and rewards, surgical operations, and various deprivations, he created artificially psychotic dogs, submissive dogs, suicidal dogs, obedient dogs, tense dogs, and dogs that had nervous breakdowns during thunderstorms.

While he worked away at his dogs, he was watched by politicians and the secret police. His techniques for influencing dogs were to be used ruthlessly on humans. Pavlov, a gentle and decent man, would have been horrified to see the inhuman results of his work. He believed that one day human behavior could be manipulated wisely for the betterment of all mankind. He could not foresee brainwashing, which became a policy of his government, or the world outlined by the American behaviorist, B. F. Skinner, in which the masses would be conditioned to labor for the common good.

Meanwhile, through all this, Fabre was watching every move of his wasps and bees, following them to their nests, opening the nests to observe mating habits and egg laying and the relationships between the sexes, assembling data unparalleled in the history of entomology. The trouble with Fabre's work, other scientists believed, was that it involved lowly insects against which no measure of human behavior was relevant. Fabre followed bees while Pavlov cut holes in dogs to see how their digestive systems worked. And Pavlov got the headlines.

The nineteenth-century notion that by interfering with the essential processes of the body a scientist could redirect behavior in predictable patterns has never lost favor. The great Canadian neurosurgeon, Wilder Penfield, exemplified this most dramatically with his mapping of the human brain during the 1930's. He could prod one part of a patient's brain with an electrical current and cause him to recall long-forgotten incidents from his childhood. He could make a patient say "Yes" or "No," or raise an arm, or cry about a lost love, or feel a sea breeze blowing in his face when he had not been to the seashore in half a century. Such work, akin to the dramatic experiments of Pavlov, gave science confidence that the secrets of life were within reach. Penfield, however, became appalled, as he grew older, to discover that the deeper he probed into the human brain, the less he understood it. He realized that he stood at the entrance to an immense labyrinth where a man could easily become lost.

Fabre and Pavlov set the study of behavior along divergent tracks. Whereas Fabre always insisted that it was impossible to know any part of an insect until the whole insect was thoroughly examined, it is significant that Pavlov never concerned himself with the mental adjustment of his dogs, only with their physical responses. This left students of behavior with the dilemma of which road to follow. The problem persisted clear into the 1930's. In the United States, behavior was lumped with psychology, where research could be confined to the laboratory and to the production and analysis of masses of data.

In Europe, however, the influence of Fabre was sending more and more workers into the field to study behavior. Fabre had seen how wasps would sting their prey accurately in the nerve ganglia to paralyze rather than kill, allowing them to provide fresh food for their young. He argued that this indicated an intelligence that was irreconcilable with any theory of the animal as a simple machine. In the end, he even came to doubt some of Darwin's main theses on evolution, so meticulous and detailed had been his personal evaluation of insect behavior.

The opposing behavior study groups had practically no communication with one another; the split continued almost into the middle of the twentieth century. The roots of both groups went deep. For Americans, the main thesis of John Locke, the British philosopher, was attractive. He had argued that every newborn child was like an unmarked wax tablet awaiting environmental imprinting. The idea that the New World could form a fresh kind of human personality was beautifully consistent with the Founding Fathers' view of America; it confirmed the philosophy of the Bill of Rights, equality, and opportunity for all.

Still later came the Freudian revolution, which gave the twentieth century a philosophic base, emphasizing the crucial influence of instinctual drives, and the importance of infant sexuality. Alfred Adler, Freud's one-time associate, disagreed with him, insisting that the motivation dominating man was his striving for perfection in the face of his own inferiority. Adler brought his school of individual psychology to the United States in 1935, but before his arrival, John B. Watson had already opened the way for the principles of American behaviorism. Watson, like Pavlov, believed that man had a limited number of innate reflexes which could be stimulated to respond predictably.

The European behaviorists were always more Darwinian. Rather than search for factual and analytical answers, they espoused the belief that every individual possessed his own unique inheritance. Life, therefore, tended to be directed by genes—not by an environmental imprint on that blank wax tablet that the Americans believed in so profoundly.

The two schools came into open conflict when the Watsonians, who believed that a stimulus could be fabricated to elicit a given response and that a dependable response could be expected from a certain stimulus, collided with the theories of the German psychologist Max Wertheimer. By 1912, Wertheimer was speculating that extremely complicated patterns of behavior appeared quite abruptly, seemingly independent of any stimulus or experience that might have preceded them. It was as though some secret signal were animating the living organism, sending it on its way, obedient to messages buried deeply in the past.

Gradually, however, the two schools have slowly come together. Each has realized that the other has something to offer. In fact, neither can manage without the other. Both the laboratory analyst and the Fabre-like field worker are essential. Behavior is certainly based on genes inherited at birth. It is also certain that the unfolding environment overlays this behavior with other influences, with the need to adapt, and in some instances, with the need to cause drastic changes in behavior and even in physical form.

A certain kind of grayish-colored moth, common in the midlands of England, clung to its habitat as the Industrial Revolution surrounded it in the 18th and 19th centuries. Its gray wings blended against the speckled bark of trees in its natural surroundings. Yet as the forests were felled and the grime and soot of industrialization spread across the heartland of England, the moth seemed destined to be wiped out by its enemies, for it was hopelessly conspicuous against the sooty backgrounds. But the moth adapted to the new conditions. It became the color of soot and grime. At the same time, those moths that were born into unpolluted environments maintained their light gray wings.

Behavioral adaption can be even faster than that of the moths. Some muskrats colonized an island off the northeastern American coast. Their natural behavior was to build lodges, half-sunken houses in ponds and at the edges of lakes, for shelter and protection. But such protection was unnecessary on the island, for the muskrats' traditional predators—foxes, minks, cats—were absent from the island. Almost immediately they stopped building lodges. In the same area, when early settlers began pirating the eggs of herring gulls from their easily reached ground nests, the herring gulls made an instantaneous adaptation; they built their nests in trees. The tree-builders survived at the expense of those that did not adapt. Tree nesting continued for years after the egg-gathering expeditions from the mainland had stopped. If this behavioral pattern had continued for perhaps only a few hundred years, a separate race of herring gulls might have developed, and then possibly a separate species with its nest-building skills greatly refined.

The legacy of Fabre was finally recognized in the mid-twentieth century. Thousands of behavioral students, using his techniques of animal study, are now exploring almost every environment. An early leader of the modern studies, Konrad Lorenz, stood neck-deep in a Bavarian lake to become as friendly as possible with a flock of graylag geese. From his field experiments and from a matching laboratory program, Lorenz developed a host of fascinating theories and evidence about behavior. He showed that the young goose usually is imprinted by the first object it sees. This, of course, should be its mother, which it will then blindly follow until it reaches independence. But Lorenz demonstrated that it might just as well be a vacuum cleaner, a wheelbarrow, or Lorenz himself. During his lifetime of experimentation, he became a surrogate mother for many hundreds of these graylag geese.

Behaviorists gradually discovered that they must melt into the environment, sharing it with the creatures under study. C. R. Carpenter, of the Yerkes Laboratory of Primate Biology, went to Barro Colorado Island in the Panama Canal Zone to study clans of howling monkeys, a task that took him to the tops of the tallest trees in the rain forest where he could watch his subjects eyeball-to-eyeball. For days he lived in the world of the howling monkeys, and his account of their social interactions became a classic in its field.

Fraser Darling stalked the red deer of the Scottish highlands. Leslie Tuck, of the Canadian Wildlife Service, scaled cliffs in the Subarctic so that he could live and observe at the same level as the millions of murres, common circumpolar sea birds. Theodore C. Schneirla, of the American Museum of Natural History, crawled on hands and knees through Panamanian jungle undergrowth to follow the army ants that were his life study. David Mech of Purdue University studied wolves on Isle Royale in Lake Superior and found that members of the pack cooperated in their search for prey. Mech tracked the wolves by airplane. A Dutch researcher, Hans Kruuk, followed hyenas in East Africa in high-speed night rides in a Land Rover. Valerius Geist, of the University of Calgary in Alberta, scaled the Rockies to study his subjects, bighorn sheep.

Perhaps no man better exemplifies the importance and ingenuity of animal behavior studies than George Schaller, a young American of German descent. He has that gift, unusual among scientists, of being both an analyst and a humanist. He established his reputation with a magnificent study of the mountain gorilla of west-central Africa, a research project of such definition, style, and care that it upset nearly all the old preconceptions about the animal.

While Schaller carefully excluded the human element from his own work, he began to suspect, when he turned to study the lion of the Serengeti Plains of Tanzania, that the examination of animal and man must eventually come together. He found his lions interacting with practically all the other major animals as he built up a complex picture of life on the plains. He discovered that in his lifetime he would never get a full picture of what was really happening. The more he studied the effects of his lions on the zebras they hunted, the more he discovered that the zebras were less influenced by lions than by internal parasites. The parasites seemed to operate independently. But they too had their own predators which arbitrated their numbers, thus affecting the zebras and influencing the well-being of the lions. He realized that the lion had little effect on the creatures it hunted. The one thousand lions under study did not eat more than two per cent of all the animals available on the plains in any given year.

He found that lions, far from being feared and efficient hunters, were lazy and clumsy. They missed more kills than they achieved. Their hunting strategies were frequently ineffective and ill conceived. They ate their own cubs or, carelessly rolling on them crushed them to death. Male lions were not cooperative—a basic trait of lionesses.

Schaller's work invites new studies which, he says, will take far longer than his own life span. He says we may never know more than a tiny part of the secret lives of the animals he is studying, and he has come to understand the awesome complexity of all life systems.

It is difficult, if not impossible, to compare the facts about one species with the data on another. Karl von Frisch, working with honeybees in Munich, discovered that when a hive was artificially overcrowded, the bees solved the problem efficiently by creating new colonies. When John B. Calhoun worked with rats in the slum areas of Baltimore, he found that in an overcrowded situation they reacted very much like the people crammed into the slums around them. But while rats may resemble people, bees do not resemble rats. The behavior of each tends to be specific to that species. Nevertheless, scientists like Konrad Lorenz *(On Aggression)*, Robert Ardrey *(The Territorial Imperative)*, and Desmond Morris *(The Naked Ape)* have outlined many of the obvious similarities between man and animals.

While no miraculous insight has yet been gained into either human or animal behavior, the work is going ahead on a broad front. The web of life is spun into billions of force lines radiating in every direction. These lines contain signals and memories, adaptations and coordinates, color messages and intuitions. All of this effort blends into one mighty objective: the struggle to survive and live as long and as fruitfully as possible. When some of these secret animal signals are better understood, man himself may benefit in ways he cannot now imagine.

The Work of the Senses

Photographs by:
P.J.K. Burton/Natural Science Photos, 29
René Catala/Aquarium de Nouméa, 4
D. Clyne/Bruce Coleman, Inc., 20
Thase Daniel, 36
Edward R. Degginger, 2, 11, 37, 39
Jack Drafahl, Jr./Sea Library, 24
R.D. Estes, 3
B. Evans/Sea Library, 22
Keith Gillett/Tom Stack & Associates, 7-9
J. Grossauer/ZEFA, 21
Clem Haagner/Bruce Coleman, Inc., 28
Grant Heilman, 16
David Hughes, 17
James K. Morgan/National Audubon Society, 34
Ernst Müller/Roebild, 26
Oxford Scientific Films, 27
Klaus Paysan, 15
Bucky Reeves/National Audubon Society, 33
Edward S. Ross, 12, 31, 32
Leonard Lee Rue III, 30
F. Sauer/ZEFA, 1, 18, 38
G. Sirena/ZEFA, 25
R.A. Steinbrecht, 13, 14
Karl H. Switak, 35
Ron and Valerie Taylor, 23
Constance Warner, 5, 6, 10, 19

When Aristotle named the five senses of sight, smell, taste, touch, and hearing, his list was pretty much complete. He described the function of the senses, but he knew little of their scope. The sense of touch alone involves many elements—the position of the body, its movement, the measurement of weight, the reactions of the joints, the tendons and the muscles. Touch is also involved with balance, dizziness, nausea, hunger, sexual appetite, and pain.

Even with our well-developed sight, and our excellent understanding of colors, we do not necessarily see what other creatures see. Fruit flies and honeybees receive light in short wave lengths which we cannot see at all. We think we hear well, and know the sounds of the world around us, but this is not true. A man cannot hear the fifty-thousand-cycles-per-second screams of bats circling over his head, much less pick up the echo of the screams as they bounce back from a strand of cotton stretched across the bats' flight path.

To survive, the animal must be sensitive to the physics of its world, and this involves a panoply of innumerable influences. The animal must be able to receive and measure vibrations, to have an accurate picture of the kinds of pressure being applied to individual parts of its body as well as the pressure present in the air itself. It must be capable of gauging temperature, and different levels of light. The tension of muscles must be monitored while cells of the inner ear register balance, and the sensory apparatus interprets what the eyes see.

The processing of signals within the sensory network may be simple or complex. When a cheetah sees a group of grazing gazelles, its eyes transmit detailed information about the herd. Perhaps the sex of a potential victim is recorded, as well as its size, its state of health, its readiness to flee or fight, and its position in the hierarchical order of the group.

Sometimes the spinal cord, not the brain, is the receiver of information from outside, and transmits it to create immediate muscle response. When a child touches a red-hot stove, no processing, or thinking, is needed. The cheetah may take an hour to make its decision to charge. The child withdraws its burned finger instantly.

The work of the senses goes on in a labyrinth that has been only partially charted. Animal behaviorists agree that the most exciting and revealing discoveries still lie in the future.

1

The compound eyes of a female horsefly (1) are marked with bands of bright color. Unlike the insects, spiders have small, simple eyes. The eight eyes of a spider (2) help her to avoid predators. The large eyes of two lesser African bush babies, or pottos, reflect the rays of a flashlight (3) because a layer of cells behind the retina of the potto's eye acts as a mirror. The small blue eyes of scallops (4) are somewhat similar in structure to human eyes. A scallop, however, has between thirty and forty eyes. The binocular eyes of the fishing owl of central Africa (5-6) are protected behind tough, nictitating (winking) membranes. This "third eyelid" keeps the owl's eyes moist and clean and protects them when the bird pounces for prey. Two sharp eyes on long, flexible stalks (7-9) peer from the open shell of a red-mouthed stromb on Australia's Great Barrier Reef.

Vision

2

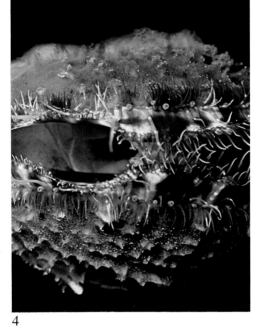

4

7

8

9

3

5

6

Smell

The senses of smell and taste are devices that are at once long range and short range in an animal's measurement of the world around it. The forked tongue and odor-sensitive nostrils of the emerald boa of tropical American forests (10) supplement its eyes. The tongue flicks out to pick up chemical particles in the air, or from the ground, and then transfers them to special sensory cells inside the mouth. Rattlesnakes are members of the viper family and inject venom into their victims. The pygmy rattlesnake of North America (11) hunts chiefly at night, relying on chance and a keen sense of smell to find food. When it nears its prey, another trailing device is employed. The pits on each side of its head, between eyes and nostrils, are sensitive to the presence and position of warm-blooded creatures and can detect the body heat of the hunted animal. The scent organs of a Madagascar butteryfly (12) show that the sense of smell is vital to an insect's survival, and its feelers contain numerous odor receptors. The scent organs on the feathery antennae of a male silkworm moth (13) are extremely sensitive to the smell of a female and can lead him directly to her. To summon a mate, the female silkworm moth (14) raises the tip of her bulging abdomen while expanded glands release her powerful lure. The caterpillar of a swallowtail butterfly (15) reacts to the approach of danger by trying to look as formidable as possible. Two slender and bright-colored projections are thrust out from behind its head. When the caterpillar turns into a butterfly, it will be equipped with a tubular organ designed to suck nectar. The swallowtail keeps this tube curled until it smells the right food flower. Then, it unrolls the tube, and stiffens it like a straw. The soft, flexible skin around the five teeth and central mouth

10

11

12

13

14

15

16

of a purple sea urchin (16) provides the prickly marine creature with information about the texture and chemical features of its food. Sea urchins are relatives of the starfish, and their needle-like spines are sometimes poisonous. The African flap-eared chameleon (17) is quick with his long tongue when catching a grass-hopper. The slender mouth parts of a hover fly (18) extend to reach the nectar in a flower. The long tongue of a hovering humming bird (19) licks the nectar from the edges of its slender beak. Hummingbirds need an enormous supply of nectar to satisfy their extraordinary energy demands and are the most efficient bird collectors of the sweet, sticky substance. The long, sucking tube of the privet hawkmoth (20) is uncoiled to probe deep into a flower for nectar. There are one hundred and forty thousand moth and butterfly species; only the beetles are more numerous.

Tasting

17

18

19

20

21

Touch

Touch is such a vital part of animal behavior, that some creatures organize their behavior almost totally on its impact. Butterflies brush antennae and "read" messages from touch. Seabirds touch beaks and so help cement bonds between them. The octopus, although it has highly developed eyes, depends on touch—as do many other animals—to learn how to avoid dangers and enemies. An octopus attacking a crab, in a laboratory experiment, quickly learned to avoid the crab if it was equipped with an electric charge. This octopus (21) uses its eight sucker-studded arms to seize and hold prey, court potential mates, and polish the gelatinous masses in which its eggs developed. Two young elephant seals (22) rub against each other, reassuring themselves they are among their own kind. The tentacles of the sea anemone (23) distinguish between food and non-food mainly by identifying

chemical substances. If the object turns out to be food, it is injected by the myriad needles covering each tentacle, and its paralyzed body is eaten whole. The feather feet of anchored barnacles (24), which smother many shoreline rocks, reach out to catch small, drifting animals which are then pulled in and eaten. Touch is extremely important among larger animals, particularly for lions (25) where, rubbing faces, a lioness and a lion perform a pair-bonding gesture common among pride animals. Among primates, touch is vital to the development of the young. Contact with the body of its mother helps a young orangutan to develop normally (26). These arboreal apes are playful youngsters but become moodier and more solitary with maturity. Touch helps a calico spider (27) to distinguish between the web vibrations of struggling prey, an entangled leaf, or a timid male seeking to mate with her.

21

22

23

24

25

26

27

23

Hearing

Hearing and balancing have a distant evolutionary connection. The ear was developed, according to scientific theory, first by fishes as an organ to help them balance themselves. If the inner ear of humans is upset in any way, then the sense of balance is also upset, sometimes so seriously that normal walking is impossible. But the ears of fish and the ears of the higher vertebrates today have hardly any points of similarity. The American bullfrog (30) can measure eight inches from nose to vent and weigh up to one pound. Its eardrums are extraordinarily large and are located just above the corners of its wide mouth. During the mating season, deep-throated roars rise from any pond occupied by these creatures. The bullfrog's hearing is rather primitive, at least compared with that of a bat. A small bat's big ears (29) hear sounds we cannot detect. The bat sends out sounds in the range of one hun-

28

29

31

32

30

24

33

dred thousand air vibrations per second. For prey animals on the African plains, interpreting sounds is essential to survival. A shy doe steinbok (28) points her ears in one direction, and then in another, to pick up danger signals before she ventures further into the undergrowth. The zebra (33) has a sharp sense of smell and keen hearing. Among the insects, only grasshoppers, locusts, cicadas, crickets, and most moths have organs resembling eardrums. These hearing devices are never placed in ears on the head, but rather on the legs or abdomen. The male ''bladder grasshopper'' (31) produces mating calls by rubbing the rough inner surfaces of its rear legs against their file-like edges. The grasshopper amplifies these sounds through two air sacs in its distended abdomen. A male cricket (32) claims territory, or attracts a mate, by rubbing its two wings against each other to produce its individual sound. Along with scent, sound is one of the most important of all the signaling systems. A sense of balance is also so essential that many creatures go to their deaths the moment they lose it. Specialized hooves give a bighorn sheep (34) secure footing as it climbs a sheer cliff wall. The bottoms of the sharp-edged hooves are concave, with soft pads at the center, so that they get a suction-like grip on the bare rocks as the bighorn makes its steep ascent. The gerenuk (36) is an African antelope which lives in arid territory and has a taste for the leaves of acacia trees. Its elongated hooves are pointed and triangular and help the gerenuk to keep its balance while it stands on two feet to feed. A young adult tree frog (35) is secure despite its awkward position on a slender branch in the Australian bush. All its long toes have sticky tips which stick to the branch.

Balancing

34

36

35

37

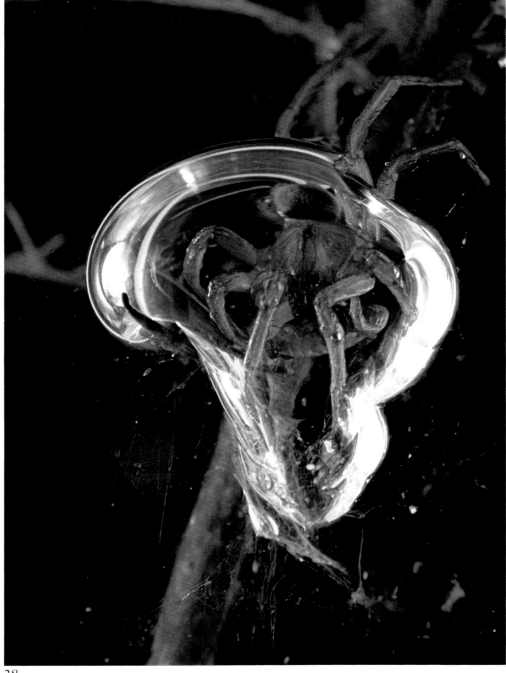

38

To be able to survive in alien environments, animals have developed many ingenious methods, particularly in their techniques of breathing. Seals can hold their breath for an hour and dive down a thousand feet. Some crabs can "breathe" for weeks without water in their gill cavities. The diving beetle (37, 39) has a gill which enables it to breathe under water. It surfaces to replenish a supply of nitrogen; all of the oxygen it needs, however, it can get out of the water through its gill. The bubble, the result of exhalation, also serves the important function of helping to balance the insect while it swims. The aquatic spider (38) has a gill similar to the diving beetle's; only the side product is not just a bubble but an enveloping diving bell.

39

The owl arrives silently at a branch where it can overlook half an acre of bushes and open field. There it is accustomed to catching rodents, katydids, moths, and beetles. Its eyes are so large they cannot swivel in their sockets; the owl must turn its head to change its view. Yet its sight is superlatively sensitive. The bird sits motionless for more than sixty seconds, watching intently, then rises slightly and crouches as it gets a fix on some slight movement in the grass that might mean food. Its wings spread soundlessly and it makes a short flight before dropping to the ground. It is not sure of the identity of its victim but it knows that the surreptitious movement means something is there. The scurrying of a desperate mouse into the grass is picked up instantly by the owl's acute sense of hearing. As it attacks and catches the mouse, special eyelids are drawn down over both eyes like protective glasses. Its sensory network has cooperated so perfectly that it does not actually need to see at the moment of the kill.

The senses are part of a natural computer where the secrets of life are stored. Here is a beginning point for the study of behavior. Here are clues to how we function on this planet. Here are hints of the great potentialities that lie dormant inside every one of us. The senses of animals and men are indeed like computers. But they are so much greater than anything conceived or built by humans that no valid comparison is possible. This sensory world contains a great reservoir of the unknown, a flood of facts indicating the subtlety, sensitivity, and power of the senses. We find butterflies that are able to navigate faultlessly for thousands of miles; dolphins that come when we whistle; marine mammals that can communicate for miles underwater with what may be language; birds that can correctly read and understand the positions of the stars; bats that use sonar; bees that are sensitive to ultraviolet rays.

Each organism responds to interacting messages generated electrochemically by the sensory cells, which dictate directions to the contracting and expanding muscles, to the tiny, beating, hairlike cilia, to the secreting and discharging glands. Each living thing is at once a receptor, a conductor, and an effector; all three functions operate smoothly together to keep the animal in balance with its world and to make it instantly ready to cope with most changes.

The senses "program" so much behavior that animals may appear to be nothing more than puppets incapable of taking independent action. Although the senses appear to be authoritarian, in fact they are merely detecting and reporting all the information necessary for an animal's survival. The sense network's "authority" comes from the genes. Man himself lives at least partially in thralldom to his senses and the genetic instructions of his central nervous system. Thus, it is difficult for him to disobey the multitude of orders, directions, alarms, warnings, and suggestions that are made to him thousands of times every minute of every day.

He stands under a low-flying jet, his ears aching with the roar of its engines, airborne vibrations impinging on his bare skin as they pass through his touch receptors. Yet a moment later, with the plane gone, he hears the keening cry of a gull, listens to the susurration of waves touching the sandy shores before him, smells the salt tang of the sea and distinguishes it from the smell of seaweed.

By supreme concentration, the man might have been able to thrust the awful roar of the jet into the background of his sensory reception and hear, beyond it, the quiet cry of the gull, since he is not a total victim of his senses. But most animals must respond automatically to the orders of the sensory network, and the lower down we go in the scale of life, the more reflexive becomes the response of the animal to the commands of the network. If we wish to understand the significance of the senses in higher animals, it is necessary for us to have some knowledge of the nature and scope of this sensory imprisonment in the lower organisms.

The exquisite sensitivity of some receptor cells, for example, allows them to record changes even in the molecular environment.

Our man-made computers cannot work without programming, without input. The senses, too, are useless unless they receive information from the outside world. They must be activated; they must receive signals.

This system can be put to work in a multitude of ways. One of the most effective is radiant energy from the stars and sun. It is energy from the sun, 93 million miles away, that keeps the earth in a temperature range tolerable to life, creates winds, evaporates fresh water from the seas, nourishes green plants, and then all other kinds of life. Seasonal cycles, as well as the change from day to night, supply a stimulus that keeps the behavior of animals synchronized with the growth of green plants. This great output of energy flows through all the arteries and capillaries of our life systems and animates the lowliest of creatures in the depths of the sea or concealed on the bottom of a permanently sunless canyon.

Radiant energy, in the form of heat or light, is a communicator that reaches the blind just as effectively as it does the sighted. The foraging earthworm, working in total darkness, quickly flees if it is exposed to light in the violet-blue-green-yellow-orange of the visible spectrum. The worm, however, is blind to red. Red light is equally ineffective in stimulating the compound eyes of a honeybee. If the insect visits flowers we see as red, it does so because they also reflect the ultraviolet from sunlight. Ultraviolet is not only visible to the bee but provides a separate, bright color in its spectrum.

The toad, sitting absolutely still on its stone in an almost pitch-black night, can be detected by its enemies because it is emanating radiant energy. The rattlesnake and other pit vipers do not need to see such victims. They are equipped with one of the more extraordinary sensory devices possessed by any animal. A conical pit is located on either side of the reptile's face between its eye and nostril. These pit organs enable the creature to receive delicate gradations of heat through 150,000 nerve cells lining the pits. These heat-sensitive "eyes" can fix the position and proximity of the snake's prey so that any rattlesnake advancing across the sand toward a bird or mouse is not fooled by the camouflage of sand-colored feathers or fur. The pit organs instantly detect the presence of the prey, and the rattlesnake strikes with deadly accuracy, even in total darkness.

It is almost certain that some insects use infrared, or heat, to locate members of their own species and the plants on which they feed. It is possible that some male moths can find their way "blindly" to females whose temperatures have been raised by recent flights. When Philip S. Callahan, an American scientist, experimented with this notion, his white-lined sphinx moths flew to the mouth of his infrared transmitter—and took up mating postures on it.

The labyrinth of the senses offers so many confusing facts, however, that any novice researcher hoping to concoct a grand truth must finally realize that nothing about the sensory apparatus is straightforward. We may think that sight, hearing, taste, touch, and smell are simple functions, but whatever we can observe is only a tiny part of the total performance.

Obviously, taste has a purpose; many living things react the same way to various sugars or poisonous substances. The bitter flavor of some insects discourages others from eating them, and the North American opossum seems to be one of the least favorite meals of meat-eaters for much the same reason. Thrushlike birds are less popular with cats than finchlike birds. Yet taste apparently does not deter the great horned owl from eating skunk flesh impregnated with repellent spray fluid.

Our own sense of taste is supplemented by touch more than we generally suppose. We distinguish between water and tasteless mineral oil more by feel than any other way. The oil seems slippery, the water does not. But can we identify water otherwise? Many animals behave as though they could, long before they swallow enough to test whether the liquid will relieve an inner sense of thirst.

Scientists generally insist that water lacks both taste and odor. Yet man and other mammals have a basis for detecting water, if only because a sip of it washes away saliva from the taste buds—and saliva contains a measurable amount of salt.

There is no known reason why pure water should have a sweet taste for human beings who have just eaten artichokes. Nor has anyone explained why a person who has just rinsed his mouth with a mild solution of citric acid also gets a sweet sensation from pure water. Other contradictions abound. Researchers have found that salt put on one side of the human tongue enhances the sensation of sweetness on the other side, while a drop of a sour substance, such as lemon juice, increases the tongue's sensitivity to both sweetness and saltiness.

More understandable is the reaction of South African gold-mine workers who were given unsalted water at the end of their working shift and immediately became nauseated. Their bodies *knew* that they needed salted water because large quantities of sodium chloride had been lost in sweat. The senses can accurately measure the loss of any vital chemical in the body and by a complex network of signals order its replacement, sometimes at practically any cost. Thus, the captive animal whose diet is deficient may eat its own feces or its mate or young in a desperate effort to satisfy what its sense network tells it is needed.

Many grazing animals, from kangaroos and rabbits to deer and antelope, get little sodium in their diet, though the sodium salts are vital to their survival and must be replaced when they are lost in sweat or urine or feces. An animal deficient in salt must find a place where the earth or the rocks on it contain soluble compounds of sodium—the well-known salt lick. But the senses of these grazers give the animals some alternatives if the correct sodium chloride is not available. They will take sodium bicarbonate or baking soda, borate or borax, sodium carbonate or washing soda, sodium nitrate or sodium sulphate, demonstrating the flexibility of the sense network in a crisis.

Only one flavor truly unites animals in a single response—bitterness. The special taste buds at the back of the human tongue define it and warn man that the substance is likely to be injurious, a signal he shares with most other mammals in a small communion of the taste experience.

But taste is a minor sense and its ambiguities are simple compared with the workings of the other senses.

29

Touch involves a more intricate signal system, enabling animals and man to hold their positions in a competitive environment. Around the base of each hair almost anywhere on the human body sensory cells report if the hair is moved, acting as simple antennae to tell us what is happening beyond the body and giving us an early warning of coming events.

As though aware how frequently survival depends upon the readiness of touch cells to respond, most animals groom themselves or one another, keeping the body surface clean. A mammal uses its tongue and teeth, and sometimes its claws as well, to comb and sanitize its fur. A bird spends hours each day preening its feathers with its beak. An insect wipes its eyes and feelers, then rubs the wiping legs together to rid itself of dust particles. Many of the bees thriftily gather together the pollen dust from their bodies and carry it home as nutritious food.

A stream of information flows constantly into the brain from muscles, tendons, joints, bones, and limbs, but we are still not certain how alert these tactile senses remain when an animal sleeps. Human responses to touch vary greatly during the hours of rest. A person who reacts violently while awake to having a feather stroked across the bare sole of a foot may, in light sleep, merely draw back the foot. The latter response is unconscious and unremembered, and in deep sleep the reaction vanishes entirely. Perhaps nonhuman animals never relax so completely unless they are in hibernation. Their tactile and auditory senses seem to continue in full operation when they are asleep.

Among insects, sensitivity to touch is refined to the point where the sensors, in effect, measure gravity, and so judge the weight of a load or report when the creature is hanging upside down. Web spiders use their legs to monitor the strands of their web and can determine where the trapped insect is located in the web by the lines that are most resistant to their pulling.

Signals pass through the great open expanses of air and water to provide long-range communication, but short-range transmission may be just as subtle, through touch, as the chemical network of communication. (Many small creatures go through their lives guided largely by their sense of touch.) The tomcat kicks the rump of the female when he wants to mate, stimulating her to greater readiness. The snail, usually hermaphroditic, begins its complex mating movements after being touched in certain places by its sexual partner. The female stickleback, a small freshwater fish, cannot lay her eggs unless she is touched on the tail by the male. The young baboon leans backward in apparent ecstasy, eyes closed, body relaxed, while a comrade systematically grooms the hair on its belly. The female digger wasp can turn her bee prey around and sting it to death only if it "feels" right in her grip.

Most fish need only two senses to function: smell and touch. Catfish and codfish use a sensory apparatus contained in cells on the soft barbels that droop below their mouths like whiskers. These cells report whether suitable food is available on the bottom where they are prowling. The sensitivity of these organs is indicated by the effect marine pollution has had on them. According to John Bardach and his co-workers at the University of Michigan, catfish will die of hunger in polluted water because they simply cannot locate food. New England lobsters starved when petroleum spills impeded their sense of smell and prevented them from finding food.

Many senses are so delicate and so subtle that they are presently beyond human measurement. Whirligig beetles are equipped with a kind of submarine sonar with which their feelers, just touching the surface of the water, register their own movements' vibrations bouncing off rocks, pieces of wood, and other inanimate objects. Water bugs of one species pick up vibrations through the hairs on their legs and quickly and accurately locate the source of the ripples, usually an edible insect.

Because our senses function perfectly most of the time, we take our responses to the world for granted. We stand upright, and that is normal. But it is only "normal" because special organs measure the direction of gravity and control the stability of the body. This control is exercised by otoliths in the inner ear. These small, hard bodies are carried on a bed of sensory hairs layered by mucus. When the body is tilted, the hairs are bent. Then, with the otoliths stimulating the receptor cells, and with the cerebellum controlling the operation, a message goes out to correct the tilt of the body.

Some aquatic insects carry pockets of air which are held in place by waterproof hairs on either side of their bodies. Bubbles increase the pressure on the hairs if the creature's body is tilted, thus giving it an instant warning of its own imbalance.

The higher forms of life receive so many sensory messages every second that they can afford to have minor malfunctions since their other senses quickly take up the job of reporting the information necessary for survival.

But in the simpler forms of life, minor distortions of the communication system may be disastrous. Many crustaceans get their sense of balance from a grain of sand that they pick up each time they molt their shells. A nineteenth century zoologist, Alois Kreidl, put shrimp into an aquarium containing iron filings. Each shrimp put a grain of iron into its gravity organ, instead of a speck of sand, when it molted. When the iron was magnetized, the shrimp swam upside down.

Sound waves carried through air and water usually contain too little energy to affect our tactile senses. Instead, sound receptors in special organs are sensitive to these waves, although animals without ears, like snakes and salamanders, can receive vibrations through the earth on which they lie. Man uses sound to communicate more complex signals than birds, whales, dolphins, or bats, but he is limited to an extremely narrow range of hearing. He can hear the cries of migrating geese keeping the flock together, or the spring chorus of frog sounds that help to bring many frogs into breeding condition simultaneously. He can hear birds singing to warn intruders or to attract females, and he can hear the night range–finder calls of petrels—swallowlike sea birds that yodel above the dark trees of offshore islands to their mates who guide them to underground burrows. Man may watch and listen as a gull recognizes his mate's distinctive call from among hundreds of thousands of other birds and goes to her unerringly. But a rat can signal another rat right under a man's nose and he will not hear the cry. The signal, also unheard by the household cat, the family dog, the canary, is an ultrasonic, high-pitched scream that vibrates in the air at more than 24,000 cycles per second. Man can only hear sounds up to 15,000 cycles per second. The high-pitched cries of whales and bats use what amounts to sonar and radar; they bounce their calls off distant objects to avoid obstacles and to find food and mates and shelter.

If we could eavesdrop on all the sounds of animal communication, we might be deafened by the great mixture of screams and howls and chirps. Occasionally, however, accidental discoveries give us some insight into this world of secret voices.

On a summer evening in 1956, Kenneth D. Roeder, professor of physiology at Tufts University in Massachusetts, was giving an outdoor party when one of his guests idly ran his damp finger around the rim of his glass, causing it to ring. Dozens of moths that had been fluttering around the Chinese lanterns instantly dropped to the ground, apparently dead. But in a few moments they became active and took off.

Dr. Roeder was fascinated by this phenomenon and with his colleague, Asher Treat, of City College of New York, he began an investigation. They soon discovered that moths of many different species could receive and react most specifically to a wide variety of musical sounds. At first, this information was merely baffling, and appeared to have no useful purpose. But as they went deeper into the problem, the scientists found that the moths' highest sensitivity was to ultrasonic sounds in the range transmitted by bats.

The months had "learned," in the vast reaches of evolutionary time, to "hear" and react to the bat sound system that was being used to locate them. To pick up such high-pitched sounds, the moths had developed between thorax and abdomen a fine membrane that covered a minute air space and contained three tiny sensory nerve-cell groups. When the bats cried out, the sounds vibrated the membrane and instantly caused danger signals to be transmitted to the brain of the insect. Each of the three nerve-cell groups has a different role to play. The first group of cells is like an early warning system, which transmits regular impulses that apparently warn the moth of approaching danger and give it enough information about the peril to take evasive action even before the bat has become aware of the moth's existence. The moment that the bat is within twenty feet of the moth, the second group of cells in the trio sends out urgent messages calling for evasive action. The moth has now been informed that the bat is directed to the kill. The situation is so dangerous that some species of moth fold their wings and drop to the ground, apparently dead. The third sensory-cell group transmits a signal which has not yet been interpreted by man but which is obviously essential to the survival of the moth.

Though the sensoring devices of the moth appear to be elaborate, the bat's own system is even more complex and involved. If a bat is released into a large room divided in half by a net with one small hole cut into it, the creature immediately heads for the hole. It produces a scream that vibrates 110,000 times per second and "scans" the entire substance of the net, instantly finding the hold and bouncing the message back to the bat. The bat "sees" more accurately with its ears than most creatures do with their eyes.

But the process is not that simple. When sensing an obstruction or possible food, some bats twist their ears back and forth five or six times every second to pick up echoes from almost every direction. They are scanning their environment without actually being able to see it. The scanning is so accurate that these bats can perceive themselves exactly in the space they occupy, an extraordinary feat of a complex nervous system. In addition to having detailed knowledge of its own position in its aerial environment, the bat stores in its memory bank a precise summary of its linear world. It can fly blind, working totally from memory, without bothering to use its sound system. The performance is so precise that a bat accustomed to flying through a maze of obstructions will continue to do so long after the obstructions are removed. Conversely, if an unexpected obstacle is raised in its path along a route that it has been flying for months, it may well crash into the foreign object.

Fish-eating bats defy simple explanation. They skim the surface of the water and suddenly drop to snag a fish with their feet. Most airborne signals known to man are deflected by water. The bat has so perfected its system of detection that it outperforms man's electronic equipment.

The dolphin's scanning system is the marine equivalent of the bat's. Cruising underwater at fifteen knots, the dolphin sends out a series of ticking noises that bounce off objects ahead and furnish a profile of its world. But, again, this is no simple signal. The dolphin transmits on a broad-band wavelength that includes both low and ultra-high waves. When it is bearing down on a victim at a speed of about twenty knots, the dolphin constantly checks its bearings by oscillating its head ten degrees to right and left. The low tones in its transmission go deep into the ocean and give the dolphin a broad-spectrum picture of what lies below. At the high, or ultrasonic, range of its transmission, it receives a close-up picture of the sea. Its high-pitched clicking noises are capable of cutting through all the interposing sounds that make the underwater world so noisy.

With this dual range-sensing system, the dolphin can ignore the myriad small creatures whose activities might blur its total picture, were it dependent only on its long-range sonar. The two extremes of sound transmission give the dolphin information so accurate that it can swim at high speed in pitch darkness through a hole just big enough to admit its body. It can detect its victims in water cloudy with mud, and it can tell the difference between a hole that is truly open and one that is blocked with clear plastic. The dolphin's sonar devices are selective enough for it to "see" the difference between mackerel and herring, between tuna and eel, between seaweed and dense masses of plankton.

An American zoologist, John C. Lilly, has theorized that dolphins come close to speech as we understand it. He has recorded them conducting what he believes were long underwater "conversations" in sounds made up of a sequence of drones, squeaks, bleats, whines, and howls. When dolphins encounter unexpected obstacles in their path they may withdraw to "confer" with clicking sounds, and then dispatch a scout to check out the nature of the impediment. When it has been determined that the obstacle is harmless, the dolphins "consult" again before moving en masse past the strange object.

Like heat sensitivity, taste, and hearing, vision works differently for different animals. The hawk can see the flight of a nestling bird hundreds of yards away and so be informed that there may be other helpless nestlings nearby. The much more simple-brained frog, sitting on its leaf, is surrounded by a host of apparently confusing details: soft movements in nearby reeds, the chuckle of water disturbed by fish, butterflies resting on branches, caterpillars clinging to stalks. But this does not confuse the frog; its nervous system apparently is capable of blanking out all objects that are motionless, thus enabling it to select moving prey with great accuracy.

Jerome Lettvin, of the Massachusetts Institute of Technology, and his associates tapped the optic nerve of a frog and thus picked up signals leading from frog eye to frog brain. He amplified the signals electronically and conducted them onto an oscilloscope, giving them shape on the screen. He was able to get some idea of the flow of the frog's information system. He actually saw how the frog measured the size of a bug coming within reach of its flashing tongue. Before the frog's sticky tongue was triggered into action, the victim had to be moving. At the University of Connecticut, psychologists Walter and Francis Kaess took advantage of the fact that most animals with eyes react quickly to a moving object. To save themselves from hand-feeding their hungry toads with wriggling mealworms, they linked an electric motor to a lazy Susan turntable and stuck bug-sized bits of hamburger around the rim. As the turntable revolved, the toads noticed the moving meat, crept over and snapped up the food. Some toads, in fact, climbed onto the lazy Susan and turned with it.

When Niko Tinbergen, the eminent Dutch scientist, was working with grayling butterflies to measure their vision, he was surprised to find out that the butterflies reacted not only to exact copies of their own species but also to a series of aberrant forms invented by Tinbergen. Although the female grayling is gray, he found that black was a more provocative color to the males and that they were super-stimulated by dummies more than four times the size of a normal female. Even the shape of the dummies did not seem to matter much to the males, so long as they were big, dark, dancing, and near. The "recognition" of the female grayling by the male, therefore, was based on some other criteria not known to man.

It occurred to Tinbergen that it was the quantity of stimulation that was important. As the experiments continued, however, he began to realize that no one factor stimulated the males; instead, it was the collective impact of many different kinds of stimulation that aroused them. In the end, he discovered that the quality of the male's response was governed almost entirely by what it was doing at the time of the stimulation and not so much by what the female was doing. The male's internal environment—not the outside world—seemed to determine what would bring a response from his sensory network.

Animals communicate visually in ways we are only just beginning to perceive. Fireflies, flashing their lights in a kind of Morse code, indicate their position, their sex, and their readiness to mate. So do creatures in the deepest recesses of the ocean, where pitch blackness rules the environment 365 days of the year. Apparently many of them flash complex information from one to another to attract mates and to warn of the approach of danger.

Of all the senses, that of smell is perhaps the longest-range communicator among animals. A specific scent, released into the night air, can travel for miles downwind to reach the sense receptors of a male insect. An odor borne for hundreds of miles in river water alerts waiting eels that have come from the depths of the ocean to a resting place in an estuary. The odor apparently tells them that this is the river they must ascend.

Even among simpler animals, an acute sense of smell—and the use of odors as signaling devices—is frequently well developed. A Trinidad butterfly displays bundles of hairs on its abdomen, greatly increasing its size, when it wishes to mate. The hairs radiate a pervasive, musky smell as the male butterfly brushes them over the female's antennae to induce her to alight. The powerful persuader can also work in reverse. The moment the male is attacked or picked up, its abdominal hairs splay and the chemical is released to repel its enemy.

Water is the ideal medium for carrying a multitude of signals over great distances. Whelks suck in water, and smell it to make a chemical analysis of the world around them. Once they have oriented themselves, their sense organs can report the position from which a particular scent is moving. The shark follows the scent of blood on an injured fish, and the tiny, air-breathing snails of the genus *Physa* know which way a member of their own species is traveling when they reach the scent trail of its slime. Indeed, the sense of smell and the capacity to create odors may well be one of the great unlocked secrets of the sea; scents may guide creatures for thousands of miles through a world that appears to be trackless.

On land, the network of smells is more ephemeral but no less complex. The dog crossing a crowded city street, following the scent of one person or thing is demonstrating the great power to discriminate among scents that is present in nearly all animals. The black fly in the Canadian north woods can smell carbon dioxide from forty feet* or more away and so locate an animal sleeping in deep cover. The snake protrudes its forked tongue, and molecules carrying scent stick to its wet surface, where they wait for transfer to olfactory pits in the roof of the reptile's mouth. Because birds are such supreme flight specialists, they have little need for a sense of smell. The vulture's keen eyes, for instance, can recognize at great distances the large flies that usually precede it to its victims. But the flightless kiwi uses touch and scent to find worms and beetles when it probes its long beak into the soil of its native New Zealand forests. Its nostrils are near the tip of its beak, and the kiwi will do almost anything to avoid using this long and seemingly powerful weapon for defense or for attack.

And yet many flies have surprisingly sensitive responses to scents. For years, students of sea birds have been puzzled by the ability of the various species to navigate in total darkness, in dense night fog, or in severe gales, to find their nesting islands, but a graduate student at the University of Wisconsin has recently demonstrated that the Leach's petrel, the Wilson's petrel, and two kinds of shearwaters all have strongly developed senses of smell. Because the burrows of these seabirds are impregnated with a musky perfume that even human beings can detect, it is probable that the birds can locate their islands at night by picking up and following the individual scents that they have placed in their burrows. So precise is this sense of smell that the petrel or shearwater can drop down through thick shrubs to the burrow-riddled ground. And the sense of smell is so individual that the bird can find its way through a maze of passages, guided by the smell of the nesting material that came from its burrow alone.

Seen from afar, the navigation of the Leach's petrel seems nothing short of miraculous as its dusky-colored body drops clumsily down through the mist-drenched trees of its island to crash among the thick undergrowth on the ground. The petrel is a diurnal bird, and its night vision does not appear to be any better than that of a human being. But its infallible navigation shows man how one sense can be refined to its optimum. The bird's behavior is actually part of an "action chain" in which a series of responses, or signals, combine to lead it to its destination. How the petrel comes to its offshore position where it can smell its island is not known, but more than just a sense of smell is involved.

When the female digger wasp is hunting for bees—her favorite prey— she flies among plants until she finds one. She may not actually "see" the bee because she positions herself downwind from the bee where she can receive the bee's scent, which triggers the attack. Once the wasp has the bee in her grip, she operates beyond her sense of smell. The bee must "feel right"—feel like a bee, that is—or the wasp is incapable of making the move to kill and carry off her victim.

The intricacy of sense perception is at once the marvel and the contradiction of animal life, leading us to believe, on the one hand, that they are possessed of supernatural powers, and on the other, to wonder why they appear to be so stupid. The explanation is that unless all the sense receptors secure their signals in perfect balance and according to a fixed sequence and timing, the behavior of the animal is likely to break down completely.

A marching column of driver ants, its route abruptly broken by the scuff of an elephant's foot, mills about on either side of the break. With their scent trail gone, the ants are virtually helpless and can only reconnect their line of march if one ant accidentally meets another, picks up the common scent, and unites the two bands again.

Specialization of the senses has quite extraordinary repercussions in animals having other well-developed senses. A large, meat-eating beetle, the *Dytiscus marginalis,* is the terror of the pond as it pursues tadpoles, fish, worms, and large insects that have fallen into the water. But this beetle frequently ignores succulent and easy-to-catch prey. The beetle can see very well, but it makes no attempt to kill. Yet when the same beetle scents a victim, even one that is well concealed in leaves or buried in mud, it goes straight to its target and makes a kill.

Though the beetle "sees" its victims, it does not truly see them, because somewhere in its sensing system the information that the creature has picked up visually has been cut out, or sidetracked, so that it will not act. Neurophysiologists call this phenomenon "gating." Its purpose is not totally understood. The beetle apparently hunts better by smelling its prey than by sighting it. A cat with electrodes buried in a nerve center behind its ear can be recorded by a scientist as it registers the ticking of a metronome. But when the cat sees a mouse, the nerve center that was registering the metronome suddenly blocks out the sound, or at least stops recording it. The sight of the mouse is the more important experience. For the beetle, the power of scent is all-important. Other animals share this reliance on one sense over the rest.

The starfish that inches its way along the sandy bottom of the ocean in search of shellfish *apparently* locates buried clams by smell. Certainly, when it reaches a buried shellfish, it digs down efficiently and soon stretches out its arms to envelop the shellfish and pull it open. Yet something more than a sense of smell may be involved. It has been found that clams are sensitive to radioactive rays and x-rays, which cause them to dig deeper into the sand. The starfish may have some unique sense which allows it to locate its victim, while the shellfish, in defense, may have developed a capacity to detect and foil the sense.

The capacity to smell explains neatly many mysteries of animal behavior, perhaps too neatly. The ichneumon fly locates the correct host on which to lay its egg even though the victim, the larva of another insect, is hidden in a tunnel or in the trunk of a tree. The fly runs back and forth across the bark of the tree, suddenly stops to adjust its position, and then drives its tiny bore into the wood like a drill. A probe finds the body of the larva and deposits a tiny egg on it. What is even more remarkable is that the ichneumon lays its egg only on the larva of a specific fly. It correctly identifies its victim without ever seeing it. Is this scenting—or is it some as yet undiscovered sense?

There is much more to a sense of smell than the mere ability to identify odors. One scientist, Adolph Butenandt, has isolated the chemical compound of the attractive scent emitted by female silkworm moths to lure males. He has found that it has an apparently simple molecular structure and is a doubly saturated alcohol. But the male silkworm moth is so accurate in his identification of the formula, C_{16}-H_{30}-O, that if a single atom is misplaced the altered scent will fail to attract him. This discovery led another scientist, Dietrich Schneider, to suggest that the sense of smell may not be involved at all; identification may depend on the *shape* of the scent molecule. The silkworm moth may *feel* rather than *smell* the scent.

In whatever way it is done, signaling by scent is part of the "language" of animals, and amazingly complex directions can be sent from communicator to responder. Female gypsy moths and silkworm moths exude from special glands social hormones, or pheromones, "lures" that bring the males to them. In the complex world of the ant and the termite, social hormones also play a vital role in the structuring of the society and may be able to transmit a great amount of information; they certainly allow ants to follow their own trails and permit them to share the trails with other ants whose scents they recognize and understand as not being hostile. For the ants, scent is the infallible guide to a hostile intruder, and it is killed immediately. Conversely, scent is used by their enemies to penetrate their societies by imitating their own odors to gain entry. Scent may even tell the ants where the best hunting is located, how many victims are available, and what sort they are.

In the sea the power of smell becomes greatly magnified, because odors travel such great distances. The smells of the sea perform a unique function in a world where microscopic creatures, floating for hundreds or thousands of miles, must be sensitive to signals that tell them when they have reached the place where they will breed, or grow to maturity, or transform themselves into new forms of life. The signaling system of chemical smells carried in the currents must guide many billions of lives without man's knowledge, though he gets odd hints occasionally. After floating in the sea's plankton in larval form, the lowly barnacle eventually drops to the bottom, becomes fixed for life, and grows its shell. Barnacles are able to gather themselves in groups and they can do this, apparently, because the adults already fixed in place give off a distinctive scent which presumably helps other barnacles to find their new homes.

Water is a great medium for transmitting information from one creature to another, often in surprising ways. Many marine creatures are masters at using electricity to navigate, to track down their prey, and even to intimidate or kill their enemies. The sea lampreys of the Great Lakes puzzled scientists for a long time because they could find fish in dark and turbid waters, attach themselves to their victims, and eventually destroy them. The lamprey can transmit from special muscles in its head a series of electrical impulses, which are reflected or absorbed by objects in its path. The electric signal guides the lamprey toward the object, which the fish can then investigate by touch and taste. Its system of electric sensing means that it can function perfectly without using its eyes.

Electricity is especially effective when creatures cannot smell, hear, or see their prey. The African soft-finned fish and its relatives probe in mud or in holes with a long proboscis, which sends out weak pulses of electricity that are relayed to special sense organs in their heads and tell the fish whether the object they have found is worth eating.

A group of South American fish, unrelated either to the lampreys or the soft-finned fish, use electrically charged tails for roughly the same purpose. The best known of these is the electric eel of the Amazon and Orinoco rivers. It navigates smoothly through turbid water to locate with its electric pulses both food and mates. Its eyes appear beady and prominent only when it is young, and by the time it has attained a length of three feet, it is blind. Scientists suspect that its eyes deteriorate because they are repeatedly damaged by the electric pulses from other fish of its species, since a mature electric eel can deliver a brief pulse almost double the strength of that needed to power an electric stove.

Charles Darwin in *The Origin of Species* said that his theory of natural selection could not account for fish with electric organs. He could not bring them together as being inheritors of some common ancestry. Only very recently, when experimenters began to use delicate measuring instruments, was it discovered that most fish use some form of electric sensing apparatus.

Since the use of electricity by animals was for so long unexplained, might creatures have perfected even more sophisticated means of using fundamental sources of energy without our being aware of it? Daniel S. Lehrman, director of the department of animal behavior at Rutgers University, has discovered that the female ring-necked dove's ovaries and oviducts enlarge at the sight of a cock dove, but that the enlargement is much less if the male has been castrated. At Cambridge University, Robert A. Hinde has reported that the hen canary can be persuaded to begin nest building at any time of the year if she is injected with estrogen, but Hinde's experiment does not encompass the entire nest-building process. The canary must eventually finish her grass nest with feathers, and this action cannot be stimulated by experimental estrogen injections. The impetus is external, and several factors are involved. The nest grass scratches her when she sits on it; she develops a brood patch—a bare spot of skin that stimulates her to collect feathers in readiness for the incubation of her eggs—and she loses her own feathers through the action of a hormone that is triggered by the presence of a male canary. The combination of all these influences also increases her supply of blood so that she will be better able to incubate.

The senses of animals thus reveal a network of interlocking signals that span the earth and perhaps the universe. Each creature is equipped with sensors that are part of a receiving station processing the host of stimuli coming into it. Like a highly developed computer, the sensors work so well together that they may call forth extraordinary responses to stimulus even when there is only a primitive nervous system and perhaps no brain involved at all. The freshwater hydra, a half-inch-long thread of life that fastens itself to underwater plants, is a hunter of small crustaceans and insects. It contracts instantly to a stump if touched or if flashed suddenly with light. This primitive creature seems only to be a double-walled sac with hollow tentacles somewhat like the fingers of a glove. It has a single opening that is both mouth and anus. The movements of the outer and inner bodies are coordinated by the intercommunication of cells. The outer layer is responsive to the pond world; the inside layer to the task of digestion. When the creature is hungry, the inner-layer cells induce contraction, diminishing the diameter of the body and increasing its length. When the outer cells cause contraction, the body and tentacles are pulled down to a knob until the danger has passed. When food appears, special cells eject threads, some of which inject poison into the victim, while others cling to the food until the hungry hydra can engulf it.

Despite its simplicity, the hydra can travel up to six inches a day when it is starving. It frees its sticky disk, which has fastened it to its home, and curves its body until it can put the disk in a new location. But how the brainless hydra can make a "decision" to move, to relocate, is still a mystery. One suggestion is that the cells communicate a shared hunger until a coordinated move becomes inevitable.

By following the path of signals through the receptors and sensors into the communication network inside the body, and then into the nerve nets, it is possible to get considerable insight into behavior. But the most careful examination has so far not revealed how a creature with neither a concentration of nerve cells that might be considered ganglionic, nor a brain can operate with purpose, direction, and seeming determination. The sea anemone in captivity reaches out tentacles to take small pieces of flesh and bits of filter paper soaked with beef broth. The tentacle puts each piece of "food" into its mouth and extends again for more. Once the broth has been extracted from the filter paper, the animal contracts its body and spits out the paper.

If the anemone is given broth-paper first, then ordinary unsoaked paper, it spits out the brothless paper immediately. Soon, however, the tentacle begins to discriminate and will not seize the paper without the broth at all. The sensory cells are able to elicit a different response through the nerve net. The animal is able to make a judgment about what is edible, but the capacity to do so is confined to the one tentacle. All the other tentacles blindly continue to accept brothless paper and transfer it to the mouth until they, too, receive new commands. Here, we seem to stand on the threshold of primitive learning and memory. But it does not last. Perhaps a brain is needed for that. The tentacle with the "discrimination" takes only about fifteen minutes to "forget," if it continues to receive broth-paper.

The brainless animal capable of purposive action offers us hints about the development of life and behavior in the deepest recesses of the past. The brainless hydra, a popular experimental animal, has revealed peculiar responses to light and vibration. In the laboratories of Western Reserve University, biologists headed by N. B. Rushforth have discovered that three different sets of sensitive cells in the outer layer of the animal's body detect three different stimuli: light, shaking, and an organic compound in solution (reduced glutathione). This compound escapes from the tiniest wound in a prey animal, and the cells in the hydra tentacles can measure the substance and report to the body whether food or nonfood is involved. The scientists also found that the cells that respond to food override the commands of cells responding to the danger of bright lights and shaking, and so proved that a brainless collection of cells could, by this odd kind of cooperation, arrive at an "intelligent" decision.

At the core of all this lies the chemistry of the environment. The simple capacity of the hydra to detect reduced glutathione is vital to the functioning of that creature, but other substances probably influence behavior in all animals. A tremendous variety of messages, therefore, can be encoded through slight variations in the molecular composition of the signaling compounds. Researchers at the Baylor University College of Medicine reported in 1973 that they had discovered the substances that comprise the molecules of memory in the brain of a rat trained to avoid the dark or to respond to the sound of a bell. The Baylor team extracted the material from the rat's brain, purified it, and then injected it into an untrained rat. The effect of the training seemed to be transferred, although this work has not been confirmed so far by other researchers. Behind the extraordinary capacity of animals to transmit complex information among themselves, without the help of researchers, lies the exquisite sensitivity of the receptor cell. Every cell in every body detects and responds to chemicals contained in other cells, or to those from the world outside the body. At the moment, there seems to be practically no limit to all the possible interactions within such a vast system of chemical communication. The cells in a mammary gland, for instance, respond by pouring milk into the channels that lead to the nipple, when they receive a specific hormone that is released by the pituitary gland when a baby suckles at the breast. The hormonal message received by the cells is composed of eight linked amino acids (the "building blocks" of proteins). Glutathione, which the hydra can detect in reduced form, consists of three linked amino acids. Since there are twenty different amino acids, any of the three could be replaced by one of the other amino acids that occur naturally in animal cells, making glutathione only one of 8,000 possible combinations with the same molecular form. Even more complicated is the molecule that trained rats form in the brain. It consists of eight amino acid units, making available a staggering 25,600,000 combinations of the molecular pattern. It is probable that the activated molecule is always more complex and extraordinary than the molecule that stimulates it. Scientists have learned a few combinations and these new, man-made keys may make it possible for us to test the door to the locked cell until we are able to open it enough to see into that secret room.

There is a molecular base to the entire, coordinated enterprise of life. All activity feeds back to the lowly, individual cell, which, added to millions of others, contributes to the behavior of the mass. Somewhere in the cell lies one of the ultimate secrets of life: how organic molecules can act together as though they were capable of "thinking."

Programs for Living

Photographs by:
Fran Allan/Animals Animals, 78-82
M. Chinery/Natural Science Photos, 62
Harry Ellis, 44-47
M.D. England/Bruce Coleman, Inc., 67
Jeff Foott, 84-86
Norman Myers/Bruce Coleman, Inc.,
87, 88
Oxford Scientific Films, 40-43
Carl W. Rettenmeyer, 69-71, 73, 74
Edward S. Ross, 48, 50, 54-56, 61, 63-66,
72, 77, 83
F. Sauer/ZEFA, 51, 52, 75
Robert S. Simmons, 58-60
Nicholas Smythe, 49, 57
Karl H. Switak, 53, 76
Larry West, 68

To run, or not to run? To fight, or not to fight? Each moment forces decisions which must be correct if the creature is to survive. It is possible that sometimes decisions are reached as the creature appears to weigh one possible response against another. But responses are usually made automatically, without hesitation. The inherited responses are those that enhanced the chances of survival over millions of years, through the experiences of its forebears. Thus, a bird or reptile sees a frog on a rain forest branch and does not eat the frog because its inherited memory tells it the frog is poisonous. And when moths hear the supersonic scream of an approaching bat, their inherited response is to fold their wings and tumble instantly to the ground.

This kind of fast action helps the simplest forms of life to survive. A falling shadow warns eyeless creatures that danger may be near, light-sensitive cells reporting that something has come between them and the sun. They react immediately to the message.

Along with these inherited responses, many creatures have simultaneously developed physical traits which give them even better chances of survival.

Many thousands of insects have almost uncanny resemblances to other insects—which are dangerous—or to their surroundings. They can resemble dried twigs, leaves, flowers, blobs of excrement, patches of sun and shade, reeds, or the bark of trees. Some creatures are protected by armor, as the rhinoceros is. Others have offensive weapons which range from the spines of the stonefish, and the sting of the scorpion, to the poison gland of the male platypus.

The prospective victim may contrive to behave like a dangerous or inedible creature, as well as look like one. If the attacker cannot be bluffed, or its attack turned aside, some creatures feign death, or put themselves into a motionless stance as they test the determination of the predator. But if the bluff does not work, another survival mechanism is ready for instant use—flight. Some Australian desert marsupials freeze at the sight of a hunter, but if the creature comes too close, the marsupial makes a giant leap to safety among the rocks. Faced by a pack of hunting dogs, the delicate gazelle may advance toward its deadly enemy, snorting and stamping its feet, before fleeing.

Whatever the technique, the inherited response reflects the great reaches of time in which survival is constantly tested against the pressures of experience.

Mosquitoes are born as eggs, and some species stick hundreds of eggs together to form a raft (40) which floats on the surface of a pond. The slender, conical tops of the mosquito eggs are water-repellent; the thicker bottoms are not. As a result, the egg raft rights itself immediately when it is overturned. After one or two days, the wet lower end of each egg releases a tiny, wriggling larva into the pond (41). After only a week or two, the larva attains its full size and is transformed into a floating pupa (42). Two or three days later, the adult mosquito carefully escapes from its pupal skin (43). Millions of wingless young cicada nymphs surface together in the spring, seventeen years after they have buried themselves. A young cicada emerging from the earth (44-45) climbs a tree (46) to finish its molt. Clinging to the bark (47), the insect splits the skin on its back to release a winged adult.

40

41

42

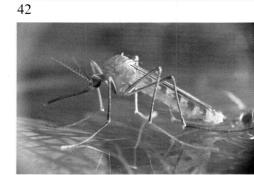

43

44

45

46

47

Camouflage

Many creatures spend their days as motionless parts of the local scenery. A camouflaged coronatus nymph (48) resembles a flower—an example of aggressive mimicry. When another insect comes to sip nectar from the "flower," the nymph will attack the insect. In India and Southeast Asia, the female of the "walking-leaf insect" (49) does very little walking. She stands motionless all day on cacao and other trees. The smaller males lead quite different lives. They are not as well camouflaged as the females and are much more active. Consequently, there are fewer of them since they are eaten with greater frequency. Only an experienced entomologist is likely to spot at first glance this perfect imitation of a green leaf as a resting katydid (50), one of the long-horned grasshoppers. Many of the measuring-worm caterpillars of geometrid moths (51) attach themselves by their hindmost legs to the branches of the trees they eat. Stretching out to look like stiff twigs, they fool hunting birds. A strand of silk spun from the mouth and attached to the branch helps to maintain the deception by keeping the caterpillar's head steady. More conspicuously placed than most of its kind, this hawkmoth (52) holds its pale antennae in an exposed position, as though it were about to fly. Among Australian geckos, this eight-inch, leaf-tailed species (53) is one of the largest. During the day, it stays motionless, its leaf-shaped tail distracting attention from the rest of its body. When night falls, the gecko's slit pupils open wide, and the lizard creeps off. The dead-leaf mantis (54) is seldom noticed in its exposed position, unless it moves suddenly to snatch an insect when it is hungry.

48

49

50

51

52

53

54

43

58

55

59

60

56

61

57

62

The bright markings of this tropical American frog of the Dendrobates genus (58) warn predators that it is poisonous. The brilliant, metallic luster of this leaf beetle (55) is instantly recognizable to beetle-hunters. They have learned that the leaf beetle is bad-tasting, and do not eat it. Another beetle, the fungus beetle (56), exhibits its own warning colors. It still contains a fungus poison derived from the food it ate in its larval stage. The gaudy black-and-yellow lubber grasshopper (57) of the American southeastern regions can be slow and conspicuous because its foul taste appeals to so few insect-eaters. The pyrgomorph grasshopper (61) of Central Africa warns predators of its offensive secretions with its bright-colored body. The ''painted frog'' of Panama (59) sits where it pleases; no predator will come near its venom-laden skin. Nor will a predator disturb this Central American

frog (60) with its colored blotches. Europe's tiger moth (62) has too many stiff bristles to be palatable to most predators.

The braconid wasp (63) in the jungles of Brazil and its reduviid mimic (66) are two totally different kinds of animals. The original is a parasitic wasp and the imitator is an assassin bug. Both show a warning coloration, alerting the predator to bad taste. This is a case of Müllerian mimicry—that is, both animals are genuinely repugnant in taste. When either bug has been tasted, both will be avoided. The lycid beetle (64) and the lycid-like moth (65) of Arizona are another instance of Mullerian mimicry. The hummingbird (67) and the hummingbird moth (68) of North America, on the other hand, represent convergent evolution, which is a case of different creatures meeting similar needs (in this case eating in flight) with similar physical appearances.

Mimicry

63

66

64

65

67

68

Defense

When quiet camouflage or brilliant mimicry fail to fool a hunter, the prospective victims frequently fall back on an inherited second line of defense to dissuade the attackers from eating them. The caterpillar of the *Leucorhampa* hawkmoth (69) not only looks like a short snake, but when it is disturbed it rears up menacingly (70-71). The threat signal from the caterpillar is so clear that the hunter tends to leave the snake-like creature alone. A tropical praying mantis (72), itself a formidable hunter with pincer-like front legs adapted to seize prey, can warn off an approaching bird by spreading its wings and legs in order to appear as large and as dangerous as possible. A plain saturnid moth (73) need only lift its front legs slightly (74) to display eyespots that will frighten off any insectivorous bird. These eyespots look very much like the real eyes of vertebrate animals and they are usually kept

concealed unless the insect is attacked. Their sudden appearance has an immediate effect on hunting birds which become so alarmed by the spots that they back off instantly. Some creatures resort to more painful forms of defense when they are attacked. A disturbed wasp (75) injects into a hunter a charge of painful venom through its extended stinger, then flies away while the hunter lies immobilized. When they are attacked, pretending to be dead works for some creatures without other defenses. The Mexican hog-nosed snake (76) compactly curls its body, with the belly up to stimulate death. The leaf beetle (77) is another creature which feigns death until its attacker moves off. This tact is so successful, apparently, that many other creatures use it, creatures not united by any kind of evolutionary relationship. The opossum plays "dead," ground-haunting birds "freeze,"

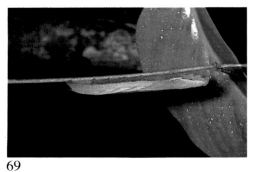

69

70

71

72

73

74

75

76

77

46

and many waterbirds blend, motionless, with their backgrounds. But many defensive and offensive measures are not inherited, or they are inherited in incomplete form. Correct behavior must therefore be learned, the inherited behavior sharpened and perfected by observation and example. Young cheetahs (78-82) stay close to their mother on the African plains and learn from her what animals to hunt and how to attack them successfully. They observe a herd of antelope from a place, usualy a termite mound, where they can get the best possible view of their prey. They must determine which animal is the least vigorous, and therefore the most likely victim. Having learned to pick the right antelope, they must combine this with the correct techniques of running the animal down, of synchronizing their blinding speed with tripping up the victim with one deft swipe of a paw.

78

79

81

82

80

Using Tools

Most animals have no trouble finding food, but to be able to eat what they find is sometimes difficult. Over the millennia of animal evolution, many creatures have developed specialized techniques to get at hard-to-reach food. Africa's chimpanzees, without doubt the most intelligent of all the apes, use a tool to find a favorite food. The chimpanzees strip slender branches of their leaves, then push these sticks into termite mounds. The chimpanzees wait a moment, then withdraw the branches. They are covered with termites which the chimpanzees greedily eat.

Some dung beetles bury their eggs in mammal manure, then shape the manure with their front legs into round balls for easy transport. The dung beetle (83) rolls the heavy manure ball to a place where the soil is soft enough to bury it. Underground, the moist dung ball is safe from the drying work of the hot sun and the desiccation effect of the wind. The egg hatches into a larva, which feeds off the dung and matures inside it. Within a few short weeks, an adult dung beetle breaks out of the crusted ball and digs its way to the surface. Sea otters living along the Pacific coast of North America hunt for sea urchins and shellfish among the kelp beds growing there. A sea otter (84-86) uses a rock as a rudimentary tool. It cracks a shellfish against the stone resting on its chest. Egyptian vultures (87-88) have learned to pick up stones in their beaks and throw them at ostrich eggs. After many failures, the stone breaks the tough and thick shell and allows the vulture to get at its nourishing contents. Tool-using is not confined to the larger, more complex animals. One of the finches of the Galapagos Islands, isolated from its mainland origins, has learned to use a thorn to hunt grubs.

83

84

85

86

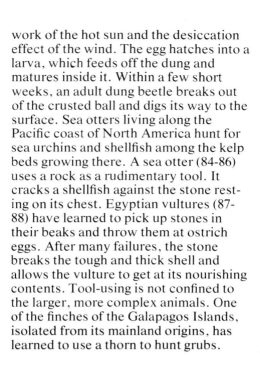

87

88

The cell lies at the base of the mystery of life; it is the great imponderable that we are only just beginning to understand. When Gregor Johann Mendel, an Austrian monk, began his experiments during the middle of the nineteenth century to prove that inheritance was governed by laws that could be understood, he pointed the way for future scientists who would eventually examine in microscopic detail the workings of the cell. Mendel theorized that there must be some kind of "unit" that transmits an immense quantity of information governing heredity. He did not have the equipment to find these units, but years after his now-famous monographs on genetics were published—and forgotten—scientists discovered that Mendel's units did, indeed, exist.

Each cell, they found, contains tiny threads inside its nucleus. These are chromosomes, and when the cell splits itself into two, the threads divide evenly, each cluster moving to an opposite end of the cell so that there will be a complete chromosome group in both cells when the separation of the original cell is finished. But when a new cell of an egg or sperm is created, only half of each group of chromosome threads goes into the new cell; thus, when the egg is fertilized by sperm, half the chromosome group comes from the female, and half from the male, making the set complete.

By the beginning of this century it was suspected that chromosomes might carry the mysterious hereditary units that Mendel had theorized. Thomas Hunt Morgan, of Columbia University, who had bred countless millions of fruit flies in his pursuit of the elusive units, discovered that each fruit fly had four pairs of chromosomes. He was convinced that he could identify the hereditary information carried by each one. For seventeen years he worked to prove that special hereditary units were indeed precisely placed on the chromosomes and that they controlled every single physical attribute of the fruit fly. The units were, of course, the genes we know so well today.

Genes dictate everything. They specify the place and size of every spot and bristle; they guide the growth of sensitive antennae, the tongue that can be curled, the compound eyes, the flexible neck, the wings and muscles to move them, the abdomen and its reproductive organs.

A single nerve cell, or neuron, placed under an electron microscope, is far more than a set of extended branches picking up excitation and conducting its output through one long nerve fiber. It is a seething mass of activity, receiving a multitude of messages over all its surface. Some stimulate it; others inhibit it. It contains a complex priority system, which is constantly being reassessed, and it is influenced by growth, nutrition, endocrine gland activity, as well as other stimuli which are not yet known. The genes in cell chromosomes specify what is possible, and a constant feedback to the nerve cell governs the sequence of responses that are appropriate to the conditions.

In the maze of an animal's nervous system, decisions are reached which are no clearer to us than they were a dozen years ago. There is no satisfactory way to explain the sequence of electrophysiological events that create "drive," "motivation," or "mood." Some behaviorists talk about separate drives for hunger, thirst, companionship, sex, sleep, and so on. They believe these drives may be responses to sensory monitors that are affected by blood sugars, blood concentration, and hormones, all working through what might be called a "comparison center," which combines information and sends out the correct instructions for behavior. It knows what *is* and what should *be*.

In most nonhuman animals, life is far too short to learn much from experience. They are utterly dependent on the information stored in their cells. They survive because they are born with a repertoire of built-in reactions. For such creatures, inheritance is vital. It governs every step of their growth and development and, indeed, their adult lives and, in many circumstances, their deaths.

This is clearly seen in the early life of the caterpillar, which is rigidly programmed as to what and where it eats, digests, and absorbs. Its heritage makes the young insect sensitive to smells and touches from vegetation which release innate and suitable responses. Heritage also patterns the caterpillar's responses to emergencies, freezing it into stillness, causing it to drop, apparently lifeless, to the ground, making it rear up menacingly, or writhe frantically. The caterpillar is programmed to respond one way to a large hunter and quite another way to a parasitic insect buzzing nearby.

The caterpillar's skin stretches tight in growth, and the immature insect begins following new behavior patterns. It sheds its old covering to reveal a stretchable skin, which may show a new combination of spots and spines. The genes dictate more mature combinations of chemical interactions. The length of the night, the humidity, the relative size of the caterpillar and its store of fat now influence the genetic system, which will precisely schedule the next skin molt and dictate when the caterpillar will go into the chrysalis stage.

But the process remains a mystery. So little is known about what abilities are innate and what are acquired by the animals themselves that man is still uncertain about how many senses creatures actually possess. They seem to demonstrate more senses than man himself knowingly uses, and the extent of their perception, with their capacity to pick up signals from the internal and external worlds, is nowhere near being fully explained.

Niko Tinbergen recounts a famous frog experiment that indicates the complexity of the problem. If a piece of skin is cut from a frog's belly and grafted to its back, the frog will scratch its belly if it is tickled on its back. The chemistry of that tiny piece of skin tells the frog's central nervous system where it *should* be located. The idea that the skin can decisively affect the behavior of an animal only makes it more difficult to understand the total function of behavior.

A psychologist at the University of Chicago, Eckhard H. Hess, once wanted to ascertain how chicks improved the accuracy of their pecking as they grew. Was it the maturing of their neuromuscular equipment? Or was it just practice? Hess goggled his chicks with prisms that deflected their vision. After days of practice, the chicks improved the "accuracy" of their pecks by clustering them, but they never mastered the prism's deflection and always pecked an inch or so to the right of the morsel of grain. A higher form of animal would quickly learn to adjust to the prism but a chick never could. Hess dubbed this "the natural unfolding of innate processes," in which some animals must continue blindly to do what they are doing and will not improve their performance much through learning by experience.

The butterfly's responses are automatic, but an individual butterfly can improve its chances of survival by finding flowers that give it a small boost of nectar to supplement the food reserve it stored while it was a caterpillar. It can sip morning dew and avoid the wind. It can respond to the sight, sound, or smell of other insects and through these signals identify potential mates and perhaps minimize crossbreeding. The pregnant female relies on scent and other signs to locate a place to deposit her eggs, repeating the inherited pattern which is shaped to benefit ensuing generations.

From the genes apparently come changes in behavior refined over eons of time. Almost every animal has its own inherited preferences that sometimes dominate its existence more than anything else. Each fall the magnificent western elk, which live in the Rocky Mountains, begin their migrations in small family groups down from the mountains to winter in the valleys. There they will live on grasses and sedges they dig from beneath the valleys' snow. In spring they slowly work their way uphill again, pacing themselves to the opening of buds of shrubs and the sprouting of leaves on low branches. They reach the mountain meadows where low vegetation, flowers and fruits seed quickly, thus giving them concentrated supplies of lime needed for the development of the young's skeletons and teeth.

East of the Rocky Mountains, the grass-eating meadow mice, or voles, mingle with the more southern cotton rats, both animals using the same tracks through the grass. But if either of them cannot survive on the existing blue grasses, they turn to the coarser quack and brome grasses to avoid starvation. Inherited preference is not as rigid as genetic endowment.

Plain mice, white-footed mice, harvest mice, and jumping mice—all seed-eaters—must eat when they find food, since they lack the fur-lined pockets to carry seeds home. But kangaroo rats, pocket mice, and spiny mice hoard their seeds. They are so pervasive they probably limit the success of seed-producing plants and, by controlling the food supply, control their own numbers. In any event, the seed-eaters rarely have population explosions.

There is tension, however, in both ways of life, with plague or starvation always a threat. To escape such penalties, a kangaroo rat native to the deserts of southeastern California follows a different inherited preference. It no longer hunts for the seeds of desert annuals, which produce only after unpredictable rains. Instead, it climbs into the shrubby salt bushes and harvests small leaves. Using its strangely flattened lower incisors, the animal shaves away and discards the salty outer cells from each leaf to reach the relatively salt-free and starchy tissue surrounding each vein. By restricting its diet to this nutritious material that is available all year, the kangaroo rat no longer competes with other seed-eating members of its kind. It has become a specialist with its own resources to exploit as it chooses.

At this level of life—among the small mammals and birds—the chances of extended survival are small. The individual barely has time to reproduce before it is dead. For example, among sparrow-sized birds that build nests on the ground or in low shrubs, about one brood in two is raised. Two hundred pairs of sparrows, producing one thousand eggs, may rear only five hundred fledglings. Of these, only two hundred and fifty will fly south in the fall and about one hundred and twenty-five will complete the round trip. Only a few of these will be ready to breed, and by the time they return for the second year, a scant sixty birds will be left of the original one thousand hatchlings.

The preferences of the individual species is aimed with pinpoint accuracy at one objective: to improve the chances of survival. But, as is always true when natural selection functions freely, a high cost is paid.

Though the song sparrow has a multitude of places to nest, by nesting on the ground it suffers the predations of ground-prowling hunters. The chipping sparrow builds its nest in the branches of small evergreens, where it is safer from the ground-hunters but more vulnerable to owls. The pine siskin builds its saddle-shaped nest higher off the ground, beyond the reach of the ground-hunters but visible to flying hunters. Nesting in a knothole may be more secure, and certainly the average age attained by hatchlings is greater, but knotholes are scarce and are frequently occupied by bees. A mated pair of birds whose inherited preference is for a knothole nest may spend so long hunting for one that they produce no eggs. Each expression of innate preference may be a gain in one direction and a loss in another. This means, though, that there is always diversity and a reduction in competition for the limited resources.

The genes of a young animal dictate a series of ironclad behavioral orders, but the creature must also learn, and often the quality of the lessons will dictate its survival. A duckling or a gosling must quickly recognize its parent and stay close to her. Its acceptance of this lead figure must be total and blind and is achieved in a peculiar way. The first moving object the youngster sees on opening its eyes after hatching is usually the mother, and the bird is programmed to follow her. It has been found, however, that if the duckling or gosling sees a football or a human being instead of its mother, it may follow the substitute.

This phenomenon among waterfowl was discovered by a German ornithologist named Oskar Heinroth, who called it *Pragung,* or coining. Twenty years later, Konrad Lorenz began his serious work on the phenomenon with geese, ducks, and jackdaws. He wanted to know the significance of such a quick attachment to the parent or substitute. Obviously it followed an inherited pattern, and Lorenz called it imprinting. The stereotype response depended on visual stimuli received by the hatchling as it emerged from the shell, and the sensitivity was lost within a day or less.

Lorenz imprinted a brood of graylag goslings to follow him everywhere. He learned some of the language of the parent bird, which he had prudently penned away from the goslings, and found he could encourage the youngsters with a goose cackle. Sounds, therefore, played a role in the imprinted behavior he was trying to understand. He also imprinted himself on young mallard ducks, but this was more complicated, since the ducklings became distressed if they did not hear an almost continuous soft signal from the mother duck or a substitute. The mallards were not interested in graylag goose calls, and Lorenz realized that the young birds knew the precise difference in the sounds.

More research has shown that while still in their eggs the young of many birds are being imprinted by the sounds of their mothers' voices. On hatching, they complete the identification of their parent by seeing her. There is, apparently, a complex and lengthy communication between mother and youngster for considerable periods before birth. An unborn baby bat may learn to recognize the ultrasonic chirps of its mother, and once it is born it will hang on the wall of the bat roost waiting for her to return from hunting. The tiny bat listens intently for her voice, and her voice alone. As soon as it hears her, the baby calls insistently. The mother knows the nuances of her youngster's voice and flies accurately through the darkness to feed it. Similar ties seem to be essential for seals and sea lions, pelicans and penguins, gulls and terns, and many other creatures. The communication system brings the hungry offspring and the parents together in the nursery area when the parents return with food to share.

Now psychologists suspect that the human infant is imprinted before birth. It is close to the rhythmic sound of its mother's heartbeat; indeed, that sound is all around it. In the isolation of a hospital nursery or an orphan's home, the playing of a record of a heartbeat at seventy-two beats per minute quiets babies and sharply reduces the frequency and duration of their crying. The same sound puts older children to sleep faster than any lullaby, but if the sound is quickened to one hundred and forty-four beats a minute, imitating the heartbeat of a frightened mother, it wakens the infants and sets them to screaming. The sound also rouses older children and makes sleep almost impossible.

As part of the behavior mechanism of goats and rats, imprinting is strengthened when the mother gives birth and sniffs and licks her offspring, thus learning the distinctive odor by which she will afterward infallibly recognize her young. Indeed, among rats, goats, and sheep smell is the only early method of identifying its mother the youngster has. A nanny goat deprived of her newborn kid before she has had any intimate acquaintance with it will reject it. A ewe will accept any youngster as her offspring as long as it has been given the odor of her own lamb. A mother rat cleans her newborn babies, but if she is prevented from doing so for twenty-four hours, she may treat her youngsters as strangers and eat them.

The behavior pattern waits within the young animal until it is signaled into action. Goslings and some pup seals do not go near the water until their mothers call them to go; once in the water, they know how to swim. The nestling flicker, wood duck, woodpecker, or cliff swallow comes out of its hole, never before having stretched its wings, and flies immediately. Young animals possess the abilities appropriate to their age. In this sense, they are the same as young children. Babies get to their feet and walk regardless of whether they have spent their first months strapped to a cradle board or free to move.

Many young creatures are keenly aware of their birthplaces. Indeed, their inherited programming demands this delayed imprinting, which may be so strong that it cannot later be erased. A lasting memory of a home area is acquired by young birds and by sea turtles soon after they reach the water. Salamanders leaving pond or stream for the first time to begin life on land may develop permanent memories of the place where they were born. The imprinting is so strong that when Victor C. Twitty, a Stanford University zoologist, watched California salamanders returning from a distance of more than three miles to breed in their natal waters, he became convinced that only a powerful odor could draw the creatures such a long distance, and he remarked that if it were not an odor, then it was not in the zoology department's domain, and might more correctly lie in the realm of the chaplain at Stanford. Animals need a varied repertoire of devices to elude their hunters. A young antelope lies still and camouflaged on the African plains, its glands not yet giving off any telltale scent to alert a hunting dog that stands nearby sniffing the air. A deer fawn lies prone and immobile from dawn to dusk in the dappled shade of the forest floor while its mother grazes elsewhere. An adult woodcock incubates her eggs in the short grass of an open field in an alder swamp, relying totally on camouflage. A dark line from the corner of her mouth to the pupil of each eye destroys the circular outline of her head and the mottled browns of her feathers make her almost indistinguishable from the dead grasses of early spring.

The programs bequeathed to animals from the great heritage of their past adroitly combine camouflage and behavior. The British entomologist A.D. Blest watched five species of closely related saturniid moths in Panama and found that the adults of two of the five species became hyperactive as soon as mating and egg-laying were finished. They were quickly hunted down by birds, reptiles, and mammals, or they died of exhaustion. But before they had bred, they had spent most of their daylight hours practically motionless and beautifully camouflaged. The other three species contained some noxious substance that repelled all insectivorous hunters; though conspicuously marked, they survived for many days after breeding.

Although the systems of survival programmed by the genes seem rigid, behavior remains highly adaptable in almost all animals, and the closer that man studies their behavior the more he is confronted with paradox. Just as he becomes certain that behavior is machinelike and programmed into the animal, he comes across instances of an almost infinite flexibility of action designed to overcome specific problems. If natural selection alters or relaxes pressures, behavior can change in a few generations. The Canada geese at Jackson Hole, Wyoming, now build their nests in trees to get away from coyotes, badgers, and weasels. But in the Arctic tundra where there are no trees, and where wolves and foxes are not so common, they nest on the ground. Each variation in behavior provides gains and losses, and when the benefits outweigh the disadvantages, the behavior becomes fixed.

To survive, the young bird must be exactly fitted into its environment. If it is raised in inaccessible rocks, in the top of a tall tree, or in a tree hole, it is feasible for the nestling to be helpless and dependent on its parents for warmth, shelter, and food. But ground-born hatchlings must be able to run, hide, and hunt for themselves almost immediately. Many ground birds, particularly the nestlings of the grouse family, react without being taught to the silhouette of a hawk gliding across the sky. It scatters them immediately into whatever cover they can find. When scientists experimented with this phenomenon by pulling a model along a high wire, they discovered that a short-necked model with a long tail invariably brought panic, but if the model were towed in the opposite direction, it was ignored. The scientists believed it was because it now resembled the outline of a harmless duck, but more thorough research revealed that the ground birds were programmed to panic not at the sight of a hawk but at the sight of the *unknown*. Since few hawks passed overhead, the hawk shape panicked them. But ducks and geese were common; the nestlings were accustomed to their shapes and thus ignored them.

The signals that trigger escape behavior are much more complex than they appear at first glance. The snort of an antelope can mobilize the flight of baboons, and a potpourri of subtle smells can send prospective victims fleeing in panic. In the sea, scallops and sand dollars "know" when a particular sea star is approaching by its odor. Swimming sea anemones in Puget Sound immediately flee when they are approached or touched by an active sea star or sea slug. Some snails can detect by scent the presence of a predatory cone shell. Each of these creatures has a different escape pattern. The scallops and sea anemones launch themselves from the sandy bottom and swim until their olfactory organs no longer detect the smell of danger. The sand dollars work their short tube feet at high speed to bury themselves in sand beyond the reach of the sea star.

Practically no animal goes out into the world completely scheduled by its inheritance. Many must face life's first problems as inefficient machines. They must be able to make many tiny adaptations to their environment. The spectacular flight of the butterfly, the wild jump of the wood duck from its high tree nest, or the quick run of the newborn antelope all tend to obscure the fact that these animals are also learning. Imprinting is instant learning, and learning is the first step in memory. Remembering must follow for the experience to have been useful. There may be a third step: forgetting, sometimes a helpful function.

The process of learning can be seen among very simple forms of animals. The slipper animalcule, a single-cell pond creature, sweeps up bacteria with its hairlike cilia and puts the food into an oral pore for ingestion. When a researcher adds powdered litmus or fine carbon granules to the animalcule's world, the tiny creature ingests these along with the bacteria. But when the foreign bodies remain insoluble, they are cast out through the animalcule's anal pore and the creature "remembers" their character so that it will reject them while accepting the bacteria. Its memory lasts only about fifteen minutes after the foreign particles are removed; it must learn all over again if it is presented with them once more.

Higher on the scale of life, the free-living flatworm, or turbellarian, can retain such learned behavior for days. Unlike the animalcule, the worm possesses a mass of nervous tissue and a cephalic ganglion. It can also swim underwater by making sinuous movements with its flat body and can explore a maze of water-filled Y-tubes, learning to improve its performance if it is rewarded with food.

The simple stimulation of being handled somehow helped the flatworms learn; similarly, the correct stimulation for higher forms is vital to the whole network of behavior. Psychologists who have worked with such higher animals as rats, monkeys, and men know the value of stimulation. When these animals are deprived of all signals from the outside world—or of any kind of contact with their species—their behavior becomes remarkably similar to that of lower animals: they regress.

Play was once thought of as a "vacuum activity" which passed the time that might have been spent in sleep, but now it has been found that learning of a latent kind grows during play. The individual coordinates muscular movements, practices, develops self-confidence, explores, and later is able to retrieve the memories of these play activities.

But memory, as we understand it, is not quite the same thing for non-human animals. The bell that Pavlov rang before feeding his experimental dogs fraudulently caused them to salivate when no food was to be given to them. Laboratory rats press a lever repeatedly as long as food appears—and continue to do so while they can hear the click of the electric relay switch in the food dispenser. The dispenser may be empty, but the rats have no way of knowing this. If the electric power is shut off, however, and the switch no longer clicks, the rats soon stop pressing.

There is a possibility that the trained animal is reacting in a context that the experimenter does not understand. Trained porpoises can be taught to bump large brown sharks, bull sharks, and lemon sharks. But Perry W. Gilbert and his associates at the Mote Marine Laboratory at Sarasota, Florida, have found that the porpoise seems unable to generalize about sharks. It learns to earn a reward for aggressive behavior toward a specific shark in a sea pen or to a number of sharks that it has met, but the porpoise is less likely to attack a strange shark, and so the notion of using trained porpoises to protect scuba divers is not too useful an idea.

At Cambridge University, Robert Hinde has suggested that discrimination and drive appear to include both the consequences of physiological conditions inside the animal and stimuli from the outside world. To complicate matters more, even the external stimuli operate in at least two ways: They help to maintain a level of alertness that keeps the animal awake, and they partly influence behavior by juggling priorities. No wonder we cannot depend on porpoises to attack strange sharks when we want them to!

And yet two other psychologists, F.J. Herrnstein and D.H. Loveland, trained pigeons to peck or not to peck at certain photographs, depending on whether the birds could detect a person in the picture. The birds learned to recognize a human form anywhere in a projected picture, whether the person was clothed or nude, adult or child, black, white, or yellow, and in almost any posture. In other words, they apparently developed insight. But such experiments give us very little idea how wild animals discriminate in natural conditions. We cannot explain satisfactorily, for example, what discriminations a pollinating insect uses in its world.

Imitation, although rare, is another way animals can learn. Fledgling birds, hearing the communicating signals of their elders, adjust their own cries to conform. Without such guidance, they might not develop the complete repertoire of their species' song. Indeed, they might be incapable of courtship and mating. Male red-winged blackbirds must hear the voices of other males every year. If this stimulation is denied them, their melodious calls quickly deteriorate into unimpressive squawks. Young European chaffinches do not develop their normal song unless they hear other chaffinches sing.

A colleague of Niko Tinbergen induced a hen canary to raise a cock bullfinch. The young bullfinch learned to sing like a canary. Later, he bred with a hen of his own species and two cocks in this brood learned the song of the canary from their father. One of these cocks was sent to an aviary two miles away and mated with a hen of his own species. The canary's song was preserved perfectly down to the great-great-grandsons of the original singer, even though none of the birds had ever heard a real canary sing.

Bird cries so well express hunger, fear, aggression, mating, that some behaviorists feel they are using the beginnings of language. Hubert and Mable Frings, of the University of Oklahoma, recorded birds' cries and found that a European gull might not respond to a call from the same species of American gull. But some British herring gulls could accurately respond to the sounds made by both American gulls and French gulls. The American gulls could react properly to the British birds but not to the European ones. Apparently the gulls had once belonged to a common group with one "language," but over the millennia, they had become separated into different groups that communicated with difficulty, or not at all, with other groups. More recently, it has been discovered that certain species of American frogs, toads, grasshoppers, and crickets have individual dialects. Midwestern creatures have a significantly different "language" from that of creatures from the southeast, but this seems to be genetic rather than imitative.

It is common enough for one animal to copy another's gesture or action, if either is part of habitual behavior. A baboon yawns, and the rest of the troop yawns in sequence. The family parakeet, seeing humans eating supper, climbs down to its seed dish to eat along with them. But when one individual conceives an utterly novel act and is then imitated by the rest of the species, we are left without adequate explanation. This has been observed in recent years. The tit is a small European bird that is very fond of fat. Apparently one bird discovered that it was possible to tear the lid off a bottle of milk to reach the layer of cream that forms at the top of unhomogenized milk. Soon other tits were following the innovator's example.

Similar novel behavior may be observed when Egyptian vultures pick up stones and smash the shells of ostrich eggs. This is an activity probably confined to a relatively small number of birds that have seen other vultures using the stones and are able to imitate them. Chimpanzees learn from older chimps how to get themselves a termite meal. They thrust the stem of a plant into the top of a termite mound and then withdraw it with a mass of furious termites fastened to the stem. Recently, in a captive troop of Japanese macaques, one of the monkeys discovered how to wash gritty soil from sweet potatoes. Soon every individual in the troop except the very oldest was cleaning its food in this new way. The digger wasp, however, inherits its ability to use a pebble as a tamping tool to close the entrance to its nest. It need not watch another wasp to know how to perform this feat.

Because of the innumerable complexities and contradictions inherent in understanding animal activity, researchers have been forced to devise ingenious experiments to expose the reality of animal motivation. For instance, does a squirrel know naturally how to open a nut and extract the meat, or must it learn this task? The problem so fascinated Irenaüs Eibl-Eibesfeldt, of the Max Planck Institute, that he reared squirrels in simulated natural surroundings, but gave them no seeds or nuts until they were adult. He discovered that the squirrels immediately recognized a hard-to-open hazelnut as food and were able to crack it open, but it took them some time to extract the meat. This skill had to be learned, apparently because the squirrels ate so many different kinds of nuts that it was inefficient or impossible for them to possess so much innate behavior. Their capacity to store nuts for the winter, however, was entirely innate.

Ghost crabs from the sandy beaches of Cape May, New Jersey, were taken into a laboratory and placed in glass dishes containing three inches of sand. They quickly learned to make new kinds of burrows there. Their initial efforts produced unsatisfactory shelters, but their work steadily improved until V-shaped tunnels were devised; these were certainly not the deep, Y-shaped burrows they usually built. The ghost crabs emerged from their new homes each evening to run not to the nearest wave to wet their gills but to a glass medicine dropper full of ice-cold sea water offered to them from the refrigerator.

The sexton beetles of North America seek the corpses of small mammals, which they bury beyond the reach of foraging ants. The beetles tend their eggs and the carcass, feeding on its body and regurgitating the meat for their young. They build a side tunnel in which their helpless offspring can pupate. A pair of sexton beetles in Haliburton County, Ontario, found a dead mouse tied by one leg to a sturdy stake. The beetles cleared the soil from under the mouse until it dangled in the air. Then one of them explored the carcass, found the string, and chewed through its multiple strands.

To use the word "ingenuity" to describe such behavior is too anthropomorphic for most scientists, but many simple creatures do make what seem to be ingenious variations in their behavior to solve problems. If the industrious dung beetle of Africa is prevented from rolling its ball of dung toward a suitable burying place, it will likely split the dung ball in half so that it will have enough strength to lift its burden up and over the obstacle. There must be a sequence of innate types of behavior that can be released in sequence to make sense of such persistent activity. Eventually, if the animal keeps trying, some novel combination may bring surprising results.

The senses constantly monitor what an animal is trying to do. No learning could be faster than sensory feedback, the fundamental behavioral control system which lets the animal quickly match its actions to a changing world. Additional learning or further changes in living conditions may stimulate the individual to use hidden behavior in his inherited repertoire. But the final test is whether the survivor, like a relay runner, can pass his experience to the next life in the genetic line.

The Mystery of Migration

Photographs by:
Glenn D. Chambers, 102
Thase Daniel, 101, 103, 104
David Hughes/Bruce Coleman, Inc., 93, 94
J. Alex Langley/DPI, 108
Tom Nebbia/DPI, 107
Oxford Scientific Films, 96
Carl W. Rettenmeyer, 95, 97, 98
Edward S. Ross, 99, 100
Ron and Valerie Taylor, 89, 90
Steven C. Wilson, 91, 92, 105, 106

The creatures of the world move about its surface with a sureness and an accuracy that must baffle the lay observer. The punctuality of their arrivals and departures is astounding. As everyone knows, the swallows arrive at Capistrano on almost exactly the same day each spring. The precision of migratory creatures is such that a seabird leaving the North Atlantic for her subpolar breeding grounds finds the exact place on one of the many thousands of cliffs where she had nested the previous year, and the year before that.

The power to navigate accurately over immense distances is not limited to the larger and more intelligent animals. Indeed, intelligence appears to have nothing to do with it. The ability to travel in the right direction at the correct time is possessed equally by small insects and giant whales. Almost all creatures, it seems, have some inborn capacity to navigate. Locusts move by the hundreds of millions, crossing continents in their everlasting search for food for their ever-growing host. Some whales migrate thousands of miles into cold waters to feed on summer upsurges of plankton.

Migration takes a multitude of forms. It may be a reflex action in response to certain subtle changes in the exterior world around the creature. It may be part of an inherited drive to go to new territory to breed or to pass the winter. It may be the animal's response to drought, flood, or wind. Ethiopian elephants, for example, once streamed south by the scores of thousands in search of better pasture when their home range was eaten out. One year, some African finches migrate hundreds of miles looking for new food supplies and breeding grounds; the next year, they do not migrate anywhere, nor do they gather together in large flocks. Caribou always migrate to their winter feeding grounds across the tundra, but the Cape buffalo of southern Africa may not move from their grazing grounds for years.

Migration appears to be an inherited imperative, but sometimes change can occur. In modern times, for instance, thousands of Canada geese have ceased to migrate north, now that a journey to nest sites is unnecessary; they have found in the south secure ponds and lakes protected by accident or by concerned nature-lovers.

Most oceanic migrations are still unknown, mysterious, and puzzling. Only in the sea is there a truly vertical migration—some creatures move hundreds of feet up an down searching for the right currents to carry them in the direction they must go, seeking shelter from the sun, or finding sustenance in its warm rays. Shrimp appear to be on an almost endless migration. But the apparent helplessness of the moving shrimp is deceptive. Each creature carries a grain of sand inside its shell, and these tiny weights allow the shrimp to maintain their balance in the swirling waters. The grain of sand is picked up by the shrimp when it sheds its shell and grows a new one. When a light is shone on the surface of the sea at night, a vast comity of rising shrimp is revealed (89). One small shrimp is isolated in the light (90) from the great corporate mass which is made up of billions of creatures.

89

90

Migrating to Breed

The salmon ranges for thousands of miles in the open sea during its growth to maturity, and then must find its way back to the river of its birth through what appears to be trackless ocean. When the call comes, some salmon begin working their way south along the coast of Greenland, heading toward European rivers. In the Pacific, they move down the coasts of western Canada and the United States to find the rivers where they were born. The salmon gather in great groups (91) while awaiting the right moment to overcome some obstacle, such as low waterfalls. Although mortality on these migrations is high, virtually nothing except death can stop a salmon until it reaches the shallows (92), where it meets a mate and they complete the act of laying and fertilizing eggs. Such feats of navigation clearly have a rich, inherited pattern. Parts of this pattern are probably derived from celestial cues, others from memory and from the "smell" and temperature of familiar waters.

The turtles of the sea are also great migrants and navigators. They have chosen the sea as their home but they must return to the land to breed. The Pacific ridley turtle (93) comes ashore in great numbers at designated times to breed on Pacific island beaches. The ridleys synchronize the time of their arrival (94), and the females dig sandy nests in which to lay their eggs while the males wait offshore. Later, freshly hatched young turtles will make their run for the sea past hunting birds and waiting crabs. If they are lucky, they will reach refuge in the turtle grass of the shallows and there will fight to survive their first risky months of life in the sea.

91

92

93

94

The Bivouac

Since migration means to move from one territory to another, the army ants of the American tropics can be called partially migratory, though their migration is unlike any other. They move almost constantly, and like soldiers on forced marches, their foraging parties push ahead in search of food while guards line the army's route to repel enemies. The soldiers (95) guarding the flanks of the moving columns of workers (96) are stimulated by chemicals secreted by the larvae which are carried by other ants. The pregnant queen must struggle to keep up with the marching columns. Once the larvae stop accepting food from the foragers and begin spinning the cocoons, they no longer secrete the substance that has stimulated the rest of the ants. The soldiers and workers form a bivouac (97), with the cocoons and the queen protected near its center. The queen is now free to lay her eggs. These

soon hatch into larvae and the stimulating secretion is again released, activating the ants. Each soldier (98) takes its position and the army ants begin to move once more on their endless trek through the tropical forests.

The migration of the North American monarch butterfly is as complex as the march of the army ants, but quite unlike that of any other insect or larger animal. Here, the inherited guidance system is a multi-staged program. On autumn evenings in many parts of North America monarchs can be seen forming great clusters on trees (99). They do not tarry long; soon after the sun has warmed the early morning air, they head south where they spend the winter in a state of partial hibernation (100). Many of the monarchs fly to Mexico from departure points in eastern Canada. Others winter in Pacific Grove, about half-way down the California coast.

95

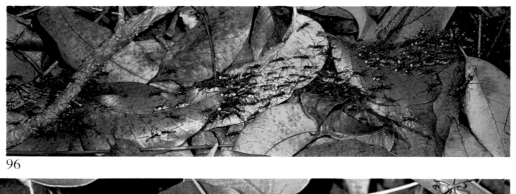

96

97

98

99

100

Seasonal Migration

The high, haunting cries of Canada geese are familiar signals of spring and fall as the birds travel between their nesting sites in the northern United States and Canada and their winter quarters in the south. Their migration routes are calculated precisely, with familiar resting and feeding stations (101) so well spaced that the birds are always roughly just below the front line and thus enjoy more or less equable weather throughout their migration. Canada geese are incomparable migrants to watch, flying by day or night in straggling V-formations (102), and giving out their companionable cries. Like other waterfowl, they move north and south along four distinct flyways—two along the coasts of the United States, one through the Mississippi valley, and the fourth close to the eastern flanks of the massive Rocky Mountain system. In sharp contrast to the orderly Canada goose is the common and ubiquitous quelea finch (103). This hardy and fecund red-billed bird is spread over wide areas of central Africa, and its populations may remain quite stable for years. But in drought, the queleas gather in awesome flocks (104) and travel hundreds of miles in search of food and breeding places. Then, in new localities, they gather into immense colonies to breed. With their populations replenished, and the drought ended, they return to their old hunting grounds.

Among land mammals caribou make some of the longest migrations on earth. Like so many other migrants, these restless northern deer (105) are driven south by their need to escape the extreme cold and winds of the far northern winter. They come south in long lines across the snow-covered tundra (106). They winter in coniferous forests, eating lichens. With the spring thaw, the caribou move north again.

101

102

103

104

105

106

Trek to New Pastures

Migration is most usually an expression of the need to move to new localities to find food and/or the best places to breed. The greatest mass migration of large animals is made by a species of wildebeest, a heavy-set African antelope also known as the blue-brindled gnu (106-107), which, when rains replenish the short sweet grasses of the great Serengeti Plains in the uplands of Tanzania, move en masse to exploit this sweet pasture. If the rains are short, and the growth of grass brief, they return immediately to the long-grass, acacia tree woodlands where they have been spending the dry season. In such a year, their breeding will be a failure. Several hundred thousand animals make the migration, which may involve more than a thousand miles of walking. When the grasses are grazed down, the herd gathers together mothers, young, "bachelors," and dominant males and moves onward.

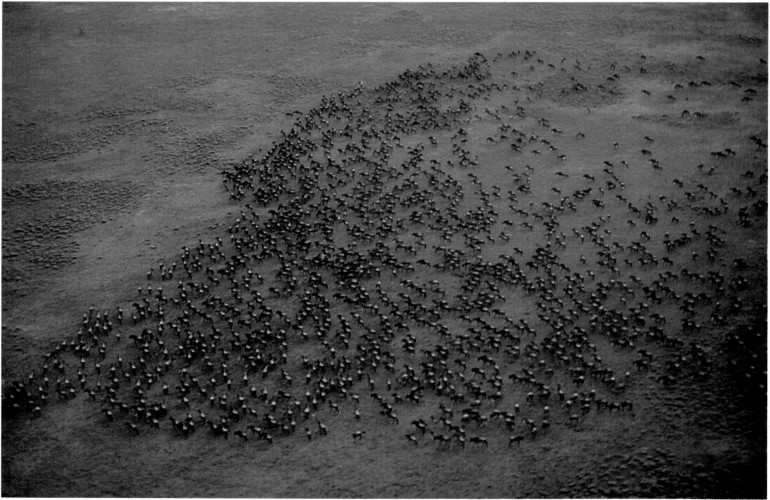

107

108

The urge to move, to change with the seasons, to pursue better weather or more abundant supplies of food, touches nearly all forms of life. Birds, fish and aquatic mammals make the most spectacular migrations because it is easier and quicker to fly or swim than it is to walk, but millions of other creatures, most notably the hoofed animals, travel great distances as part of the same cycles. Caribou leave the cold northwest American tundra around July and migrate south for up to five hundred miles, following well-established paths. Nothing stops their urge to move, and sometimes whole herds drown when they attempt to cross flood-swollen rivers. Stimulated by drought rather than by cold, African antelope and zebra move several hundred miles in a season to follow the fresh growth of succulent short grasses that quickly sprout after rain. The great mass of safari ants driving forward in long columns to search for food are, in a sense, migrating, as they dig tunnels beneath the earth to circumvent obstacles and to protect themselves against hunters. Geese migrate across the Himalaya complex of mountains at nearly 30,000 feet, a height at which human beings cannot long survive without oxygen, much less match the supreme physical exertions of the geese. Millions of other birds commonly fly at between ten and twenty thousand feet. Female Alaskan fur seals winter off the coast of California, about 3,000 miles from their summer home at the Pribilof Islands in the Bering Sea. Whales travel long distances in migrations that so far have not been accurately plotted by man, but one tagged fin whale, swimming between Antarctic and tropical waters, was found to have traveled 2,000 miles. The tiny pipistrelle, a European bat, flies about seven hundred miles between southeastern Europe and the central provinces of Russia. Migration has very little to do with physical strength; many of the most powerful animals are largely sedentary, while some of the weakest and flimsiest of creatures may migrate thousands of miles. The mass movement of monarch butterflies from Canada to Mexico has been recorded frequently, and Mexican free-tailed bats, famous for their dense occupancy of New Mexico's Carlsbad Caverns, make migratory flights of nearly one thousand miles into Mexico.

United in their urge to change their situation, the migrants move: Squid pour in toward the shores of Newfoundland from the deep Atlantic; salmon plunge up rivers throughout the northern hemisphere; turtles move up the shores of South America and across the Caribbean; individual codfish strike out from their birthplaces to swim hundreds of miles to new feeding grounds; herring engage in long, circular migrations that

The Mystery of Migration

67

cover scores of thousands of square miles of sea; the Arctic tern flies from one end of the earth to the other. The urge to migrate has the force of natural rhythms, and its power is greater than that of any individual. This may be a poetic rather than scientific truth, but the miracle of migration indeed suggests the lyric quality of animal life.

Within the mystery of migration is concealed a multitude of secrets. We do not yet fully understand, for instance, how animals manage to point themselves in the right direction. When deer mice were transported two miles from their comparatively small home territory, the tiny creatures found their way back in forty-eight hours. Some bats have returned home in twenty-four hours after having been moved twenty-eight miles. Countless cats and dogs have turned up at the front door when their owners had given them up for lost. While many of the directional aids used by animals are still unknown to us, the secrets of migration are slowly yielding to the scrutiny of animal behaviorists. The great network of signals, some from outer space, appear to stimulate creatures in different ways. In the laboratory, scientists can induce certain animals to reset their calendars to correspond to that of their artificial environment. For a long time it was presumed that temperature alone guided the migration of birds. But William Rowan, a biologist at the University of Alberta, suspected otherwise, and many years ago he began to cage juncos and crows, subjecting them to different lengths of light each day. As a result, he was able to bring some of his captive birds into breeding condition in mid-December, despite the bitterly cold Alberta weather. But when he sought to apply his experiment to other birds, he was less successful. Not all birds, it would appear, are sensitive only to the length of light in each day. Rowan eventually came to believe that many migrants have annual breeding cycles that have nothing to do with the length of the day. He was right to assume that each species has a distinctive sensitivity inherited from its forebears and passed on with mathematical accuracy.

Among many animals there seems to be some connection between the production of sex hormones and the act of migration. The growth of ovaries and testes is tied to hormone production by the pituitary gland, and certainly light stimulates the gland to release hormones. But when birds are blindfolded, the pituitary gland still reacts. Although it is responsive to light, the pituitary does not have to receive it through the creature's eyes. The gland can sense the presence of light by its action on nearby parts of the brain. As with so many other aspects of animal behavior, it is clear that the urge to migrate is a complex, interrelated series of activities based on physiological changes which occur in the creature's body and which are probably triggered by external stimuli.

When the migrants are ready to travel, however, they seem to have the capacity to measure both time and the apparent movement of the sun across the sky. Shearwaters and seals, sea turtles and salmon, refer to these inner clocks to keep themselves in position, as it were, while the earth rotates and orbits, and the sun rises and swings across the sky. For their clocks to work, they must know exactly what time of year it is, as well as the time of day. This adjustment to time and the placement of sun and stars must be accurate to the second, even by sophisticated human measurements. Because oceanic birds are so frequently driven far off course by storms, they seem to possess an especially refined capacity to navigate. They have been taken hundreds of miles beyond their home territories in random directions so that there would be no chance of their recognizing any geographical details along the way. Yet a shearwater taken from eastern America to Great Britain, where none of its kind has ever visited, invariably finds its way home, back to a single island that a human would have difficulty finding without complicated instruments.

On a smaller scale, the honeybee displays spectacular navigational skill. It returns to its hive in a straight line, a "beeline." Its accuracy is made possible because it "knows" the exact position of the sun and the time of day. Even if the bee is captured and kept in a darkened place for several hours, its navigation remains faultless upon its release because the location of the sun tells it the time of day. In fact, it does not even need sun; a large area of bright blue sky will do. The bee can make sense from sky light because its compound eye can analyze the polarization pattern and thus place the sun in its proper position. Only in dense overcast or fog does the bee become lost. On such days, it remains in the hive.

The ability to navigate is inborn. Hatchling salmon inherit a complex program of swimming directions that enable them to head in the right direction when they leave the place of their birth. This guidance system, however, is not needed when the fish return to fresh water to spawn. We know that fish are extraordinarily sensitive to scents. Salmon can follow a scent if only one molecule of the odor enters their nasal cavities. This ability enables them to pick up the scent of the river in which they were born and to follow it upstream until eventually they reach the gravel bed in which they were spawned. Presumably, the same hereditary directions are passed on to each new generation, although it is less easy to understand how the fish find their way back to their home rivers from the open sea.

The process of natural selection eliminates those young fish whose built-in guidance system fails to lead them accurately downstream, but the survivors that return to breed frequently face agonizing readjustments. Because the shallows in which they were born are influenced by storms and ice, the spawning beds sometimes disappear and the fish may be forced to choose appropriate new locations. The spot in which a salmon puts her eggs must roughly correspond to the place where she was born. If she is, in fact, passing on a fixed inner-guidance system to her offspring, she must choose a place that at least approximates her own birthplace, in order to give them a fighting chance.

Rarely is one factor influential in determining the success of a migration. It is a coalition of many different forces that results in action, and it is obvious that animals must possess other senses we may suspect but cannot yet measure. Indeed, human beings also have senses which are not fully understood. Extrasensory perception, telepathy, and other psychic powers almost certainly have their foundation in some as yet unexplained sensory system.

Behaviorists have not had difficulty proving that salmon navigate by scent and memory, or that birds and many other creatures navigate by the stars, but the European robin defies these explanations. Indeed, after a series of exhaustive experiments, Hans Georg Fromme, of the Zoological Institute of the University of Frankfurt am Main, was forced to conclude that the European robin used methods presently unknown to man. Fromme's robins demonstrated that they could navigate when the sky was overcast and when they were locked in darkened rooms. He was unable to find any outside force which could be guiding the robins. It was not a magnetic field, because he used strong magnets to disturb the earth's magnetism, and this had no effect on the robins' navigational abilities. Electromagnetic short-wave radiation reaches the earth from space in almost immeasurably small quantities of force, yet this did not appear to have any effect on the robins, either. The only positive statement Fromme could make after his intensive experiments was that when he put his robins into a concrete chamber they became disturbed and seriously disoriented. When he moved them to a steel strong-room, they became utterly disoriented. Many forces operate on earth and in the universe, such as micropulsations—extremely long waves with single vibration times of nearly two minutes. Man has difficulty measuring such forces, but perhaps the European robin does not. Many birds use the stars to navigate during migration. E. G. F. Sauer, of Freiburg, hand-reared some European warblers one spring and kept them through the summer and into the fall. At no time did he give his birds any chance to see the sun, the moon, the sky, or any concentrated source of light. When the birds normally would have begun migrating south, Sauer took them in a closed box to the Bremen planetarium. There, in the darkened auditorium, he let them see on the domed ceiling a projection of the stars as they appeared in Bremen at midnight at that time of year.

As they watched the planetarium sky, the caged birds moved to face south. Then Sauer covered the warblers with a black cloth while the projector was changed to present the same pattern of stars, but with the polestar to the east instead of the north. When the cover was removed, the birds shifted their positions and again faced the "constellations of the southern sky." It seemed clear that if they were free to travel, their intention would be to go south. When the projection was changed once more to give the warblers a view of the stars they would normally see in the territory where their kind spent the winter, the birds looked at the stars for a few minutes, and then went to sleep. They had apparently reached their destination.

Birds are particularly dependent on excellent eyesight for survival, especially during migration, and their eyes are disproportionately large for the size of their brain. Evidently it should be in their best interest to have unimpeded vision, yet a bird's eye contains a large, blood-filled object called a pecten. It has a complex shape, folded and heavily pigmented, and its presence gives the bird a substantial blind spot. The pecten apparently is not sensitive to any form of light and its role remains ambiguous.

Migrating birds carry their own fuel. On short flights they do not have to carry large quantities, but intercontinental travelers are equipped with long-range "tanks." The fuel is the fat stored in their bodies. In flight, this fat is consumed to keep them going. Its production is precisely controlled by the internal clock that tells the bird that migration time is approaching. Nearly 20 per cent of a bird's body weight may be transformed into fat in preparation. By the time the bird reaches its destination, the fat content may be down as low as 5 per cent. Fat makes an ideal fuel. It holds almost twice as much energy as the equivalent weight of carbohydrates or proteins. It can be stored over much of the bird's body, it is not an inconvenience when the creature is flying, and it is readily consumed, moving quickly from the cells into the bloodstream before it is oxidized.

The greatest fuel tanks carried by any bird are those of the migratory hummingbirds. In nonmigration seasons only 10 to 15 per cent of their body content may be fat, but as migration approaches, the fat level zooms to nearly 50 per cent. The tiny hummingbird, winging along at fifty miles an hour across the Gulf of Mexico, in a flight that may be five hundred miles nonstop, fuels itself on 1.3 grams of fat.

Many migrant creatures cover immense distances by a steady series of short flights. The purple finch makes the journey from Connecticut to New Brunswick, Canada, easily in four days, never flying more than twenty miles an hour or more than a few hours a day. The blue-winged teal flies about 122 miles a day from Quebec to make the 3,300-mile trip from Quebec to Guyana. Lesser yellowlegs fly from Massachusetts to Martinique in six days, averaging better than 300 miles a day.

The creatures move by day or night, the decision influenced by their vulnerability, their capacity to feed, their size, and their speed. Most warblers, flycatchers, and orioles move at night, but not all the small and vulnerable perching birds make this choice. Hawks and swifts, swallows and herons, ducks and geese, crows and doves usually fly by day.

The day migrations are spectacular. Thousands of crows pour along a narrow path in the sky to new winter feeding grounds. Dizzying flights of birds of prey soar over Hawk Mountain in Pennsylvania where they catch great updrafts to give them the height to continue their southern flight. The invisible night migrants enliven the skies with their penetrating, piping cries, which is probably a way to keep in contact with each other. Nearly all the migrants travel to quite specific areas, perhaps to the same field where they were born, perhaps even to the same tree or cliff. Certainly they always go to the vicinity of their birthplace. Like birds, fish usually migrate seasonally and apparently in response to changing temperatures or the desire to breed. Many freshwater fish stop eating during cold weather and drop deeper into the water where it is warmer. Some wait out the winter in the bottom mud. The cod migrates to establish spawning grounds in the springtime of the sea, and the Pacific halibut goes down about 150 fathoms in the Gulf of Alaska to spawn. From there, its young will be carried toward the coast of British Columbia; there they will mature before returning to their parents' spawning grounds to spawn themselves. Tuna migrations span entire oceans. Schools of bluefin tuna enter the Gulf Stream and the Caribbean, proceeding steadily northward until they reach the coasts of Newfoundland and Nova Scotia, where they decimate the great feeding schools of mackerel and herring.

The migration of eels is one of the most extraordinary expeditions on earth. All the eels of the Atlantic ocean, both European and American, swim to the Sargasso Sea, between the West Indies and the Azores, to spawn and die when they are between five and eight years old. Their youngsters must return to far-distant freshwater rivers, where they will mature. American eels migrate westward, taking about a year to reach their many destinations on the continent, and they are small adults when they arrive. European eels, sweeping west, then north, and then east in the grip of the great, circling current of the Atlantic, take nearly three years to reach their homes. Yet, when they arrive in Europe and the British Isles, they are the same size as their American cousins were earlier; their retarded growth prevented them from maturing fully at sea.

The pinpoint accuracy of much animal navigation makes it difficult for man to believe that creatures find their way solely by matching the time of day against the position of the sun or moon or stars. One has only to think of eels, so deep in the ocean that they move through Stygian blackness, or the Leach's petrels heading toward their islands in thick fog, to be convinced that for some creatures many other factors are involved.

Fish use electricity to hunt, to navigate, to repel enemies, and to kill their prey. Perhaps, then, there are creatures, sensitive to the magnetic fields of the earth, which carry around in their bodies some sensory counterpart of a mariner's compass. Though we may suspect that animals can measure magnetic fields, we have no clue yet as to what part of the body is used as the sensor. Researchers have established the existence of such sensors by stimulating a number of animals with artificial magnetic fields much more powerful than that which the earth naturally possesses. The mud snail, the European cockchafer beetle, and several kinds of flies, among others, have shown an extraordinary response to this laboratory-induced magnetism. But a great deal of work remains to be done before we know precisely how these creatures use the earth's magnetic field.

We do know that a sense dubbed the inner clock guides creatures to correctly timed assignations with mate, food, and place. The tiniest creatures and the biggest possess these clocks. The elephant has one, and so does the crustacean no larger than a grain of wheat. Tiny beach hoppers on the sandy coasts of the Mediterranean infallibly leap toward the water when they are disturbed. They do this even though their first leap takes them up a slope before they can go down toward the water. Their internal clocks are so sensitive that if they are taken from the southeastern coast of Italy and released on the opposite shore, they will automatically head inland, as though they were prepared to climb the Apennine Mountains in order to reach their former homes.

The internal clocks must work with absolute precision if they are to persuade a creature to leave the relative comfort of its winter home before the snow has even melted at its summer residence, perhaps thousands of miles away. Africa is the winter refuge for millions of water birds that breed in northern Europe and Asia. They come pouring into the continent by a variety of routes, some of the waders and ducks crossing the Sahara, others filtering down the western coast and then driving inland to reach the great Niger swamp. When the Niger River floods, inundating enormous areas of land before it recedes, it acts as a kind of magnetic pole, drawing fliers from every point of the compass. Other thousands of European migrants, among them eagles, kestrels, and owls, use the African plains as hunting grounds when it is winter in their summer territories. And yet, when their internal clocks tell them it is time to leave the warmth and abundance of Africa, the birds do not hesitate. While many creatures move in immediate response to the prospect of better living conditions, most migrants seem to be more influenced by their inner clocks than by expediency.

The life of any animal is greatly simplified if it can live by repetitive patterns, which the clock helps to provide. It can register the rising and setting of the moon or sun, the changing of tides, the passing of seasons. A trained observer can watch animals switching from one sensory system to another as they obey the most urgent orders being received.

Fiddler crabs, which are common along many sandy shores, come from their burrows only when the tide is low and do their foraging before the tide returns to flood them. Thus, they must work on a twenty-four-hour solar rhythm combined with a twelve-hour-and-twenty-minute tidal rhythm. To protect the crabs, the internal clock also dictates that the translucent shells on their backs will be darkest during the high tide that occurs at midnight, and palest during the low tide at noon. The clock operates perfectly while the creature is at liberty, but experiments have revealed that the efficiency of the clock depends on the constant corrections imposed upon the crab's sensory apparatus by the changing environment in which the fiddler lives.

In captivity, the crab goes on with its original activities, but the timing is off. The darkening and lightening of its shell reach their extremes a few minutes earlier each day, indicating that the sensory mechanism governing the change runs fast when it is uncorrected by the environment. The clock, deprived of its environmental stimulus, sends the crab out to look for food, not forty minutes later every day (to match the times of the low tides), but almost sixty minutes later. But the clock can be reset artificially by the experimenter. If he turns on a bright light for only five minutes each day to simulate the sun, the clock returns to normal. The clock's governing of activity can be corrected by briefly flooding the crab's laboratory environment with sea water.

Frank A. Brown Jr., of Northwestern University, has tested these rhythmic responses of animals under laboratory conditions where light and other stimuli were kept constant. His nocturnal animals—bats and mice—began each period of activity a little later; the diurnals—birds and lizards—started each day a little earlier than the previous one. The animals had no way to correct these repetitive errors since their stable environment gave them no impetus to change.

Most of these daily rhythms are inherited, but some day-loving creatures can become nocturnal when conditions demand it. This is especially true during migration when countless millions of birds, which see no better by night than men do, take to the dark air to fly to their winter or summer quarters. Perhaps they are responding to the danger of winged predators, or perhaps they find it easier to navigate at night because they are guided by the stars. The Leach's petrel comes to its breeding island at night because darkness cloaks it from its omnipresent enemy, the prowling herring gull. Beavers and coyotes teach their young to be active when mankind is not; they prefer to be abroad when the moon is full and high, a time that gives them both light and privacy.

Rhythmic responses work at every level of animal life. In the oceans, an immense up-and-down movement of creatures occurs practically every hour of the day. Before sunset, a great host begins to swim, or drift, upward to the near-surface, where feeding starts in darkness. Then, before dawn, the host descends again, sometimes going as deep as six hundred or seven hundred feet. The great marine biologist Sir Alister Hardy believes that this vertical migration has great survival value, since it gives the tiny creatures a chance to stabilize their positions during their nonfeeding daylight hours, while the currents move across the surface above them. When they return to the near-surface, they are among new masses of food, and the cost in energy is much less than would be needed for a horizontal movement to new pastures.

Even so, the effort expended is still formidable. Many tiny crustaceans swim upward for six hours—from four in the summer afternoon until ten at night—to feed. The downward journey takes another six hours, from two until eight in the morning. In winter, when the days are shorter, the crustaceans change the schedule. They begin the upward trip at two in the afternoon, and start down at four in the morning. The persistence of this effort gives some indication of the strength of the rhythmic imperative and the power of the internal clock.

Even more finely regulated is the tiny fruit fly. After the first two days of its larval development, it can set its internal clock by any kind of stimulus that resembles day, even if it lasts only five minutes. From then on, through the remainder of its larval life, through its pupal transformations and its emergence as a winged fly, it can keep track of time. This precision allows the creature to escape from its pupa and expand as a mature fly two or three hours before dawn, thus giving the young fly time to fill out its delicate wings and harden its body so that it will be in the best condition to face its first day of flight. The inner clock of the fruit fly continues to govern its behavior and will arouse it sexually at exactly the right time, which is different for each species of fly. This strict regulation prevents the fly from attempting to court or mate with a fly of another species. If the species were allowed to hybridize, it would stand to lose its genetic distinction. So the clock dictates that the sexual encounters of the different species always occur at different times in the predawn hours of the new day.

Sometimes animals use their sensory mechanisms to resist the powerful influences in their external environment. Thus, the tiny bottom fish feeding in shallow water in the track of the Labrador Current is able to keep its position for weeks against the force of the current trying to push it toward the Gulf Stream. And yet it can make complicated migrations up and down, and back and forth across the current in pursuit of moving food supplies. It is able to absorb an amazingly complex set of signals and use the information to place itself in the right position at the right time.

On the bottoms of streams, insect larvae avoid the rushing water by clinging to stones or creeping into backwaters. Fish in the stream can adjust their positions to a familiar landmark. They do this by judging the feel of water passing along the sensory lateral lines on each side of their bodies; the memory of this feeling enables them to keep their positions in the stream all night.

The sense of place is so strong among some creatures that after a thunderous flood has scoured out a rocky creek and the waters have subsided, the same water striders, back swimmers, and beetles almost immediately reappear to resume their lives. Large trout remain in the same pools year after year, unaffected by flood or drought. Foraging limpets leave the rocky depressions that fit their shells to feed on algae, sometimes traveling five feet before finding food, and yet each creature returns unerringly to the rough hollow it has scratched out of the rock.

The net-building caddis worms do not merely hold their positions in quick water; they make their living from it. These insects are common on every continent and can be seen attending to their brown nets from South America to Asia, from Europe to India. Each larva has a slender, caterpillarlike body with two strong hooked legs at its rear end and six legs near its head, which support it while it walks over algae-coated stones and through water mosses. The larva must find an unoccupied place where the current flows swiftly. Only there can it secure enough food to survive.

When it has chosen its home, the caddis worm grips the rock with its two hind claws and lets its body stretch downstream in the water's flow. Guided by the six front legs, the body swings like a pendulum while the caddis worm makes its net. The insect deftly fastens droplets of saliva linked by slender threads to the rock. These threads become as strong as silk. The first row consists of a curved series of loops attached to the rock and bowed downstream in the middle. Then the caddis worm backs up slightly and fixes another saliva drop to the center of each loop to start its second line of netting. Each row goes slightly farther at the ends than the preceding row so that it can be securely attached to the rock. The net, held taut by the force of the current, does not tear or pull loose, for it is flexible and billows. When the net is finished the caddis worm goes to the end of it and waits. Soon the net will catch a water flea or some small aquatic animal and the caddis worm will eat it.

Pond caddis worms of different species do not behave in the same way because there is no natural rush of water for them to fight against; no flow to bring them constant food and oxygen. So the pond caddis worm responds to the challenge by creating its own private current. It builds a small, portable tube for a home, in which it undulates its body so that water enters one end, passes through the caddis worm and aerates its gills, then is forced out the other end. The larva, in effect, acts as its own pump.

Some caddis worms make cases from fragments of leaves cut into rectangles and cemented to rock with viscous saliva. Some build their own "log cabins" from short lengths of twigs. Other species use sand grains or the shells of tiny snails. Charles Kingsley, who was an interested caddis-worm watcher, wrote about the creature in *The Water Babies* in 1863: "One would begin with some pebbles; then she would stick on a piece of green weed; then she found a shell, and stuck it on too; and the poor shell was alive, and did not like at all being taken to build houses with . . . then she stuck on a piece of rotten wood, then a very smart pink stone, and so on, till she was patched all over like an Irishman's coat." Caddis worms do collect a catholic selection of oddments to make their cases, but Kingsley's creature was no more extraordinary than the case builders that live in pools near Hot Springs, Arizona, and select only shiny pieces of opal for construction material.

Animals are incredibly sensitive to the stimuli that guide them to their destinations, tell them when to mate, when to defend territory, and when to seek different kinds of food. But their sensory network is not always perfect. The great locust swarms of Asia and Africa are programmed to begin long migrations into the wind. The inflexible guidance system, which the locusts submissively follow, works beautifully as long as the locusts do not stray too near sea coasts, where the prevailing winds may be driving inshore. If this happens, millions of locusts obediently turn away from the land to fly into the wind, only to find themselves lost in the infinity of the sea.

Responding to a signal, millions of young spiders climb to the summits of their refuge places and spin out their silken parachutes to ride the air currents. Presumably, they are sensitive only to a breeze which will blow them in the correct direction. But the young spiders can have no idea that large rivers or lakes may lie beneath them. Sometimes, in the late afternoon, as the steady day-wind dies away, the tiny spiderlings settle in the water, their silken parachutes near-to-invisible in the clear air, the failure of their ingenious migration complete.

But spiders have successfully colonized many parts of the world by using their parachutes for transport. Fifty years ago an expedition discovered immature jumping spiders at 22,000 feet in the Himalayas. Major R. W. G. Hingston, a naturalist who was with the expedition, assumed that the young spiders were cannibals that ate other spiderlings carried uphill by the wind, since they were living nearly 2,000 feet higher than the nearest known flowering plant upon which the insects might feed. For a long time Hingston's theory was accepted, but then it was discovered that springtail insects, or snow fleas, lived above 22,000 feet, and that these creatures were chased and eaten by tiny black spiders on bright, sunny days. The springtails lived on pollen grains and other organic material blown up to the snow fields by the wind. Microscopic algae were also found at that altitude. They had adapted to being frozen and thawed countless times. And so, in that rarefied, freezing world, the hunting sequence continued uninterrupted among the black springtails, the black spiders, and black glacier worms which are kin to earthworms. These creatures live minute by minute, so drastic is the climate. Every passing cloud lowers the temperature to a point where hunting must end immediately, and each night the creatures are frozen solid.

For all its awesome complexity and mystery, migration solves one fundamental problem of animal life. The journey to safety is a sensible response to changing conditions that would kill the migrants if they did not move. But the significant point about all this activity is its unending diversity. Each species inherits a distinctive sensitivity from its forebears and passes it on with near-perfect accuracy.

Secrets of the Standfasts

Photographs by:
Robert J. Ashworth/National Audubon
Society, 128
Fred Bruemmer, 134, 135
Jane Burton/Bruce Coleman, Inc., 112
A.J. Deane/Bruce Coleman, Inc., 109
Jack Dermid, 129
B. Evans/Sea Library, 126
Walter Ferguson, 131
David R. Gray, 116, 117, 133, 136
George Holton, 125
Aldo Margiocco, 110
Robert W. Mitchell, 137, 138, 140
Eiji Miyazawa/Black Star, 120-122
C.E. Mohr/National Audubon Society, 113
C. Allan Morgan, 118
Norman Myers/Bruce Coleman, Inc., 124
Charlie Ott/National Audubon Society,
115, 119
William Partington/National Audubon
Society, 111
Klaus Paysan/ZEFA, 123
Edward S. Ross, 132
Leonard Lee Rue III, 127
Robert S. Simmons, 139
Karl H. Switak, 130
Cyril Toker/National Audubon Society, 114

Birds, it has been said, never developed large brains because they were able to solve their problems by flying away. There was no adaptive pressure favoring intelligence; wings provided them with an all-encompassing solution. But for millions of other creatures life must be faced in the one place. Without wings, they are either too small or too weak to be able to escape danger or deprivation. They are the standfasts.

But to stay put requires qualities that are perhaps even more specialized than those possessed by the movers and migrants. These creatures use an extraordinary panoply of devices which range from camouflaging colors and the capacity to remain rigid even when a hunter is breathing into the victim's nostrils, to thick woolen coats with layers of protective fat against the cold and exceptional abilities to store water inside the body.

The standfast must make do with what there is, and therefore must be resilient. The hibernating squirrel drops its heartbeat to such a low rate that the drain on its body resources is very small. But the groundhog, whose heart does not beat so slowly, fattens itself before going to sleep and is able to use this fat to survive the winter. The pupating case of an insect, nestled in the bark of a tree, is camouflaged to escape the eager eyes of a nuthatch. The pupating insect is also protected by the chrysalid's insulating tissue so that only an eighth of an inch of bark is needed on either side of it.

The standfasts must face and survive all the extremes of the earth's climate. Tiny rodents high in the mountains where winds may gust to three hundred miles an hour in winter do not follow the larger animals down the mountain in migration but lie in deep burrows thickly lined with grasses cut during the summer. Beetles in the blazing hot dessert are not dried out and killed by the midday sun because they burrow into the relatively cool sand. Quail do not die in blizzards because they throw themselves into snowbanks and wait out the storm in relative warmth and safety. The standfasts may be conservative in their habits, but they are also well-adapted survivors.

Sleep is the greatest defense against change in the world around the standfasts. Snails (109) estivate (summer sleep) during dry weather and are thus able to survive in deserts. They also become comatose in cold weather and can be found hibernating under the ice at the bottom of ponds. Most squirrels do not go into true hibernation, but sleep in a nest (110) when excessive cold forces them to reduce their energy needs. The dormouse of Europe and Western Asia (112) is a true hibernator. It curls up for the winter among fallen leaves, its body functions profoundly changed. Like many reptiles, box turtles (111) also seek hibernation places undergound. The hibernating pygmy bat (113) takes precisely the reverse route to survival by seeking the highest possible place to fasten itself to the walls of its rock cave.

109

112

110

111

113

116

117

118

119

114

115

78

The seasons change, and the animals caught in them must change too. The most dramatic of these changes is made by those creatures which completely alter their appearance from one season to the next. By making itself look like its background, the hunting animal has a better chance of sneaking close to its prey. By the same token, a potential victim tries to become as inconspicuous as possible to avoid being detected. The little Artic fox (116) wears its dark coat in the summer when the background of its tundra world is darkish green and gray. But when snow falls, its coat becomes pure white (117) and thickens to protect it against the cold. The snowshoe hare also transforms itself according to the color of its external world, although this process is not always completely synchronized with the changing conditions in regions out of the far north. It is almost invisible set against summer growth in northern woods (114).

But at the northern end of its range where snow may persist for months among the trees where it lives, its coat becomes pure white (115), giving it a chance to elude the hunting foxes, weasels, and wolves. Color changes are not restricted to mammals. Many birds go through color changes that appear to make them less conspicuous when they are migrating, or to identify their age and their readiness to breed. The grouse-like ptarmigan (118) is almost impossible to see against a summer backdrop of rocks and low-growing plants, even when a hunter is within inches of it. The bird is even equipped with a bright patch of red on its head which resembles some of the flowers that blossom in its summer world. But once winter comes, the ptarmigan is totally transformed. It drops its summer plumage and grows white feathers (119) to match the snow. However, it retains the red patch on its head which, in an Arctic "white-out" when drifting snow makes both air and ground identically opaque, leaves the ptarmigan visible through its red spot as it speeds along in flight. Snow and cold do not suit any of the primates. Monkeys prefer the heat and humidity of the tropics and they do not have great resistance to cold. But there are some exceptions. The Japanese macaques, or snow monkeys, which thrive on the chilly mountain slopes of Japan (120), must cope with heavy snow during winters that last for months. They have made at least partial adaptation to their cold world by developing thick coats, but these are not enough to keep them warm. Most primates fear and dislike the water, but these monkeys seek out thermal pools (121-122) where they immerse themselves completely.

120

121

122

Adjusting Body Heat

Large animals which do not have an inborn way to regulate their body temperatures are often uncomfortable in heat. African hippopotamuses (123) wallow in mud. African elephants (124) wallow, and hose themselves down. In the deep Southern Hemisphere, the bull elephant seal (125) twists and turns in a long narrow puddle of mud to build up a wet earth shield against the sun. The elephant seal (126) lounges on a hot beach and throws up damp sand with its flippers and the evaporating moisture in the sand cools its body. But heat can be advantageous for the cold-blooded reptiles. Troost's red-eared slider turtle (127) basks for hours in the warm sun, and the heat seeping into its body speeds its digestion. Crocodilians specially relish sun-bathing, as demonstrated by this alligator (128) lying in a characteristic pose on a bank. Warmth stimulates reptiles.

123

124

125

127

128

126

81

130

129

131

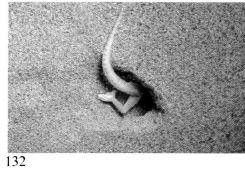

132

A combination of body and behavior changes may help animals survive, particularly in deserts and polar regions. The leopard gecko (130) has a fat tail which holds moisture and food. It can drop the tail if threatened but quickly grows a new one. Ruppell's fox (131), a desert-hunter from Algeria to Afghanistan, rids itself of surplus heat through its large ears and long tail, where blood runs close to the surface. The ears are also specially sensitive to picking up quick changes of direction made by night-hunted prey. A desert lizard burrowing into sand (132) must be buried during the heat of the day or the sun will kill it. But the capacity to dig quickly and deeply may also save it from another night hunter. A scorpion (129), like most other desert hunters, is a night creature which moves about only when temperatures are cool. Its body is made to resist the heat and is practically watertight, so that the

moisture is held inside of it. Conversely, the Arctic fox's short ears, short legs, short tail and thick fur (133) combine to help it retain a maximum amount of body heat. The thick fur of the musk ox (136) hangs down on all sides of its compact body and is practically blizzard proof. In conjunction with the animal's sedentary habits, the fur helps the musk-ox to preserve its body temperature against the most vicious outside conditions. Equally well protected is the harp seal (134), which lives inside double insulation: a blubber blanket under its skin, a dense fur coat on top. Such protection enables the seal to travel from the Arctic into temperature zone waters without stress. The walrus (135) is even more fully protected from the cold by blubber and fur and its large body mass. It has no need to come south to whelp, as the harp seals must do. Instead, it can ride out the worst of the Arctic conditions.

133

136

134

135

Living in Darkness

Remaining safely in one place does not depend entirely upon making seasonal adjustments to changes in the world around the animal. Some creatures stand fast in the dark, hiding in caves from hunters. The blind crayfish (137) is so in touch with its environment that it can feel its way with its sensitive legs and antennae along the bottom of a pitch-black cave pool, and scavenge food from the bottom. Also concealed in darkness, hundreds of bats (138) hang from the roof of a refuge cave. Their echo-bouncing cries guide their flight in the dark. A hunter surprising them in their cave would have as little chance of killing one of them there as it would outside in the night air. The blind, cave-dwelling salamander (139-140) has so far advanced its adaptation to perpetual darkness that it now has no need for body pigment or vision. In its world, darkness means nothing.

137

138

139

140

For those animals that are not programmed to migrate, keeping themselves alive during cold and food-short winters can be a severe test of their inherited capacities. Honeybees feed on stored honey and keep themselves warm in the hive. Tent caterpillars and praying mantises get through the winter as eggs laid the year before. Butterflies shelter under pieces of loose bark to survive. Woolly bear caterpillars curl up under fallen leaves to await the coming of spring, when they will make cocoons and become tiger moths. Sphinx moths survive the cold underground. The pale green luna moth gets through the winter in a loose cocoon among leaf litter. Several other species of moth survive in tough silken coverings hung in shrubs or trees. Autumn signals the traveling insects to time their development so that they will enter the most cold-resistant stage of their lives and settle in the best possible locations before the changing season brings frost and blizzard.

Most nonmigrating animals must have this capacity to live through extremes of warmth and cold. Man has never mastered the trick without use of shelter or voluminous clothing. If he had to survive as the average standfast animal does, he would be confined to a relatively small part of the earth's surface. Animals have at least two ways of handling extremes of temperature. Some must maintain an almost constant body heat to survive—they are called *homoiotherms*. Others let their body temperatures shift to match that of the outside world—and they are called *poikilotherms*. These two clumsy words more precisely define the difference between animals than do the widely used terms "warm-blooded" and "cold-blooded." Many "cold-blooded" animals, in fact, are extremely sensitive to temperature changes, and some regulate their body temperature by changing their behavior.

Two experimental psychologists at Harvard University, P. N. Rozin and Jen Maye, discovered the degree to which ordinary goldfish will respond to variations in temperature. Individual goldfish had been taught to swim part way through a one-inch hole in a Plexiglas shield and push against a lever to secure the reward of a pellet of food. The two researchers decided to use the goldfish in a temperature test by running hot water into the aquarium and giving the fish the opportunity to earn the reward of a squirt of ice water if they pressed the lever. They discovered that as soon as the temperature of the aquarium went above 91 degrees some goldfish began to push the lever. One well-trained fish, put into water of 106 degrees, a temperature at which it could not long survive, worked frantically at the lever to save itself from death. The goldfish proved themselves capable of adjusting their behavior.

Until recently, man knew very little about the changes that occurred inside the bodies of creatures as they responded to warmth and cold, but in the last ten years, radio transmitters and electric thermometers have been designed that are small enough to be concealed inside free-moving animals or attached to their bodies. This equipment has given us a new look at the internal life of the fish.

A mackerel feeding in a school is several degrees warmer than the water around it; a three-and-a-half-pound skipjack tuna may be more than 15 degrees warmer than the water, and a large bluefin tuna weighing between four and eight hundred pounds may be 36 degrees warmer than the water in which it swims. The tuna maintains an almost constant inner temperature despite the fact that its blood passes through fine filaments in the gills where it is chilled nearly to the temperature of the water. The fish stays comfortable because the blood vessels running to and from the gills pass through warm muscles near the front pair of fins. As the blood goes to the gills for aeration, it transfers its heat to the cold blood returning from the gills, so that most of the heat stays inside the body, and the fish remains warm and active despite the icy water around it. Harvard's experimental goldfish were too small to be warm-blooded in cold water. Each had only about five square inches of body surface and weighed about one ounce, not enough to conserve heat. By contrast, a giant whale weighing more than one hundred tons has nearly ten ounces of warm body for each square inch of skin, as well as a thick coat of blubber to act as insulation. Other animals also have insulation to protect themselves from the cold. Muskrats and otters and diving ducks, for instance, tolerate icy water by preventing it from reaching their skin. The water is stopped by a buoyant blanket of air trapped among fur or feathers.

Like the whales, leatherback turtles manage to keep warm because they are so big. They swim nearly 2,000 miles from their Florida nesting grounds to eat jellyfish off the eastern coast of Canada, and they keep their body temperatures around 30 degrees warmer than the water in which they are feasting. Smaller freshwater turtles cannot emulate their bigger, ocean-going cousins, so whenever it is feasible they climb out of their cold ponds and streams to bask in the sun. This not only raises their body temperature but also accelerates the chemical processes which speed their growth and reproduction, and hasten digestion and the transfer of nourishment into the yolks of their eggs.

Many animals work better within a framework of behavior in which the unknown is reduced to an absolute minimum. These creatures avoid the perils of migration and protect themselves against changed conditions by hibernating, the antithesis of migration. Hibernation demands practically no effort, and there are few surprises because the creature is well hidden and asleep. The contrast between migration and hibernation illustrates the fantastic range of animal response to the environment. While the hummingbird is making its incredible journey across the Gulf of Mexico on its tiny supply of fat fuel, hibernating creatures lie in dens, their breathing slow and uneven, their body temperatures roughly matching the temperature of the air around them. They have stored fat on their bodies too, but it will be used in a totally different way. Instead of burning the fat at panic speed as a migrating creature does, the hibernator draws on its fat like a miser. It needs little oxygen because the metabolism of its body has been lowered. The thirteen-lined squirrel, for instance, breathes between one hundred and two hundred times a minute when it is not hibernating, and its body temperature fluctuates between 90 and 106 degrees Fahrenheit. This tiny, energetic creature is transformed in hibernation. It breathes only once or twice a minute and its heart rate slumps to as low as five beats per minute. In that state the squirrel looks almost lifeless, yet like all hibernators it remains highly sensitive to sudden changes in its outside world. Were it to be thrown abruptly into the water, it would begin to swim immediately.

In hibernation all bodily functions are reduced to a flicker of normal operation. Many of the endocrine glands stop working altogether. While the physical responses may be roughly similar from animal to animal, there is nothing uniform about when they hibernate, or why. Woodchucks go into hibernation long before winter comes and while their world is still quite well stocked with food. Some animals begin hibernation before the heat of summer has waned, others wait until the first snows before they reluctantly seek shelter. Hibernation unites many creatures of totally different physiques and habits. Raccoons and snails, bears and clams, earthworms and chipmunks, spiders and turtles, are all hibernators.

In its most extreme form, hibernation involves freezing the sleeper. To prepare for this drastic change, many insects dry out their bodies before winter descends upon them. Some creatures spend the cold months trapped in a solid mass of frozen soil, and many spiders survive the winter encased in ice crystals. For digging insects like larval May beetles, ants, and termites, it is safer to burrow below the frost line to find hibernation sanctuaries. Winter sleep is by no means confined to adult creatures. Many embryos and eggs hibernate. Aphids and tent caterpillar moths lay eggs that survive the winter after the adults have died. The tent caterpillars wait in protective capsules that break open when the cells multiply in the spring.

Hibernation, the most extreme technique used by the standfasts to escape the penalties of cold, is almost exactly the same as the response—called estivation—of some animals to heat and drought. When their ponds dry up some fish survive drought by estivating in the mud, their body functions reduced as low as those of hibernating animals. Some desert snails can survive for more than a year in severe drought, their life forces drastically reduced. The larvae of some caddis flies burrow into the mud of their drying pools and remain there, estivating, until the water returns. African lungfish make tubes in the dried mud and line them with mucus; they can survive drought for weeks in these shelters.

Hibernation and estivation are not foolproof methods of survival. Although the drawing on the fat reserve is extremely slow, it is prolonged. A brown bat, for example, may lose more than 30 per cent of its weight during winter sleep. A fat ground squirrel can lose nearly 80 per cent of its weight. Hibernating mammals cannot withstand below-freezing temperatures for long. If winter cold steals into their caves, burrows, dens, or hollow trees, their breathing speeds up almost immediately and they are forced to draw more heavily on reserves of fat in an effort to raise the heat of their bodies to the normal hibernating level of 39 degrees. If they fail to do this, they die. Hibernation, in fact, is rarely a long, uninterrupted sleep. Many creatures waken to eat or drink, and some even make forays outside in the snow. Animals like chipmunks, opossums, skunks, and pocket gophers go into a deep sleep that is close to hibernation but not nearly so extreme. These creatures waken readily, while the true hibernator may take hours to recover his original body temperature.

When some moths, beetles, bees, and wasps are active, heat from their flight muscles warms the thorax region and raises the temperature of their heads and abdomens. But the same insects become helpless and torpid on cool days unless they ready themselves for flight by going through a warm-up period. The sphinx moth clings to the bark of a tree and vibrates its wings. It looks as if it were shivering, but the action is more akin to the warming of a plane's engine before takeoff. A large sphinx moth weighs between two and six grams, about the size of a small hummingbird, and like the hummingbird it can move rapidly forward and backward, or hover. It moves dexterously in front of flowers, extending a long tongue to reach the nectar, but its flight muscles must be at their proper working temperature for such maneuvering. On hot nights, when moths and environment are both at 85 degrees, the insect takes off and flies efficiently in thirty seconds. On cool nights it may warm up for ten minutes or more to raise its thorax temperature from 60 degrees to takeoff heat, which is between 85 and 100 degrees.

Not all insects require warm temperatures in order to work. Hairy-bodied bumblebees can forage under overcast skies in 40-degree weather when no other insects are abroad. The hairy covering of their bodies apparently insulates as well as fur or feathers. Physiologists believe the hair saves the bumblebee about half the energy it would otherwise use to keep its flight muscles at operating temperature, which is about 90 degrees when the bee is active and the air is no warmer than 75 degrees. Bernd Heinrich, a research entomologist at the University of California at Berkeley who has studied bees and sphinx moths, finds there is a relationship between temperature and nectar. The concentration of sugar secreted by plants varies according to not only the kind of flower and its age but also the time of day. In the cool of early morning, the fireweed, a favorite of bumblebees in the Northern Hemisphere, offers a low yield of sugar in its nectar. The bumblebee drains each flower before other competitors can get into the air. It can work at lower internal temperatures because the fireweeds grow in clusters, enabling the bee to climb from blossom to blossom, thus saving the energy that flight would require. The bee's early morning forays allow it to make up in quantity what the nectar is lacking in quality. By midafternoon the sugar concentration in the nectar rises from 50 per cent to around 80 per cent, but by that time so many insects are visiting the flowers that a single blossom holds only a small, concentrated sip for each. Natural selection keeps all these variables under constant adjustment; the insects pollinate the flowers despite harsh weather and at a minimum cost to the plant, and they are rewarded with nectar.

The adjustment of internal temperatures is critical among fragile, warm-blooded animals. Many of the nectar-sipping birds must make special adaptations to each chill night. Because they live at such elevated intensity on their high-energy diet of sugar, they continue to need energy during the night. Hummingbirds work hard to fill their crops every evening, and they begin to hunt before dawn in order to shorten their night fast. Some hummingbirds allow their body temperature to fall at night to conserve energy and to cut the rate at which they lose heat to the outside world.

Deprived briefly of its vital sugar supply, a tiny hummingbird appears paralyzed and gives no resistance when it is handled by man. In its starved state, it weights about half an ounce, but if it is given nectar or sugar, it increases its weight by 50 per cent in less than fifteen minutes, and in that time it completely recovers its mobility.

Tropical nectar-sipping fruit bats go through a daily cycle similar to the hummingbird's. Their temperatures sag toward that of the outside world during their time of rest, then their body heat is raised at dusk as they prepare for flight. The hibernating insectivorous bat changes its body temperature up and down to conform to its environment during its annual food crisis. It is not completely at the mercy of the outside world; it can make internal adjustments that are fitted to the cyclic conditions of the area in which it lives.

It is difficult for bats or hummingbirds to maintain body temperature because of their small size. But at the opposite end of the size scale, another difficulty confronts a large, cold-adapted animal when it moves into a warm environment. It must get rid of its body heat or die.

The largest of the fin-footed mammals are the elephant seals that come out of the water at Guadalupe Island, off the Mexican coast, and haul their one- to four-ton bodies up the warm beaches. The water is cold but the sun is hot, and by midmorning on a clear day the great seals are absorbing heat faster than they can dissipate it into the air and sand. At first they roll on their backs and turn their reflective bellies to the sun to reduce absorption. Then they use their flippers to throw moist sand over their bodies. This evaporation cools them for a while, but by noon the heat is too much, and the great bulls, accompanied by their harem cows, inch their way like grotesque caterpillars to the water's edge and wet themselves thoroughly. They remain there until late afternoon when the sun's heat fades and they can safely move back up the beach again.

A water-bound creature cannot make its thermal adjustment as easily as a creature free on land. The five-inch-long horned lizards of the American southwest tilt their flat, oval bodies as nearly as possible at right angles to the sun in the early morning and the late afternoon, erasing the remnants of nighttime chill, or slowing the loss of heat at the end of the day. They stay active in the morning until the ground becomes too hot; when it cools off in the evening twilight, they quickly burrow two inches down into the soil. Their eyes are closed and no visible light reaches them, yet they are sensitive to thermal clues that tell them of the passing of time.

A team of investigators at the John B. Pierce Foundation Laboratory in New Haven, Connecticut, implanted delicate electric thermometers into some blue-tongued lizards from Australia and recorded their temperatures while the animals moved. The thermometers were located under the skin, in the large intestine, and under the preoptic region of the brain. The researchers found that five cells in the brain site caused an increase in spontaneous locomotion when they were warm. Three other nearby cells responded in a similar manner to a decrease in temperature. By themselves the eight cells did not determine whether the lizard would move in the direction of a warmer or cooler place, but they did serve as a confirmation center. Messages were passed through the nervous system to them from temperature senses in the intestinal wall. Only those responses of the thermometers placed under the skin seemed unrelated to what the lizards did.

For those creatures which are defenseless and cannot build their own shelters, or which live in a world where refuges are scarce, a way must be found to help them survive. Hermit crabs are perhaps the most adept at creating homes; they seek empty snail shells to use or shelter. The largest of all hermit crabs lives on islands in the South Pacific. There, it eventually outgrows the snail shell that sheltered it while it was maturing into a coconut crab, a creature so powerful that it scales trees to nip coconuts free and then opens them with its strong claws.

These crabs show us that growth does not occur without uncertainty. When the time comes to move from an increasingly tight shell, they display a deftness and what seems like anxiety to the human imagination. The hermit uses its pincer-tipped first pair of legs to twist and turn a new-found empty shell, checking both the exterior and the opening. Once the crab makes a decision to move in, it does so at headlong speed, whisking out of its old shell and ramming its body into the new one, twisting and turning quickly to check space, fit and balance. If the new shell is not satisfactory, the hermit makes a lightning-fast decision and whips back into its old shell.

In the endless struggle for survival, portable shelters have tremendous advantages, and few creatures have had as great a success as snails, which are found all over the world. The snails have had to make many adaptations to achieve their present good fortune. The freshwater snail has become able to breathe through an opening that it can close and to use its mantle cavity as a lung. The system works well when there is ample moisture, but in dry weather the snail must withdraw completely into its shell and close the door tight. It secretes across the shell opening a film of mucus, which by sealing in the limited stocks of water it has stored, gives the creature a chance to survive.

Several snails have invaded the deserts. Some which live in the Negev near Bathsheba, Israel, have developed chalky white shells that reflect the desert sun. From April through October the members of one species lie scattered about the wind-blown soil as if they were dead. If the winter rains do not fall, they can stay motionless for as long as twenty-seven months. Members of another species cluster together on the stems of small shrubs a few inches above the ground. Only when the ground becomes thoroughly wet do these desert snails come to life; some begin hunting soon after lichens and algal cells have started to grow on the surface; others attack the new-grown foliage bursting from the bushes above.

Such desert survival is a marvelous balancing of many hidden adaptations. Knut Schmidt-Nielson and his associates at Duke University have found that these desert snails can survive when the surface temperature is nearly 160 degrees—far hotter than the snails' physical capacity to withstand. But in the last few hours before they become dormant, the shrub-prowling snails cement themselves to the stems, where the air temperature rarely reaches 130 degrees. They also enhance their resistance to heat by reflecting more than 90 per cent of the solar rays striking their white shells. The snail that estivates on the ground, oddly enough, is even cooler because, by drawing its body up into the blunt peak of its spiral shell, it leaves an air space inside the largest whorl to act as insulation. The snail's body is thus almost 20 degrees cooler than the temperature in its shadow.

Other experts on survival are the rough periwinkles, which have colonized shorelines in the Arctic and in North America as far south as New Jersey and Puget Sound. After a rain they prowl the rocks above the high-tide zone and feed on the salt-tolerant lichens the rain has softened. When the lichens dry out, the rough periwinkles cling to the rocks and exist on an inner store of nourishment and moisture. The rough periwinkle has separated itself from other periwinkles, which feed on seaweed, by retaining its fertilized eggs through all the stages of their development. Young rough periwinkles already have protective shells when they set out into the world and behave exactly as their parents do. The youngsters of the marine periwinkles, on the contrary, drift along coastal shores as swimming larvae and are eaten by many hunters. Very few survive to settle down to shoreline life in their portable shelters.

Shellfish can swim surprising distances when one considers the burden they carry. A scallop, using hydraulic jets, glides through the water as gracefully as a bird, its blue eyes gleaming. The tiny wedge clam exposes itself to the turbulent water of the breaking wave and rolls shoreward with it. At exactly the right moment, the half-inch bivalve flicks out a tongue-shaped foot and digs itself quickly out of sight among the sand grains. Before the next wave arrives, the wedge clam extends a short neck, which has a pair of openings. It inhales water with food and oxygen through one opening and discharges wastes through the other. When the wave is right, the clam pops out of the sand to ride in the direction it should go.

The bivalve mollusks of the sea maintain the flow of water that brings them oxygen and food by means of microscopic, hairlike cilia that drive water from the mantle cavity through the gills and out of the body while a fresh supply is entering. Shellfish are acutely sensitive to impurities in the water they are breathing. If sediment begins to settle around it, the creature may cease to take in water, thereby depriving itself of oxygen and food. When it is forced to move again, it collects the mud inside its mantle cavity, where muscular contractions then "spit out" the indigestible material. Razor clams are expert at this removal job. They slither up out of the mud a half inch or more, spewing out the muck, then pull themselves down until only the soft tips of their intake-outlet systems are visible.

The sensitivity of the bivalves to water quality is so great that they adjust ciliary movement and fluid intake to match seasonal changes or storms, as well as polluted water. When their activity is low, the deposit of lime around the shell rim is increased, and this gives us a record of their growth. It is so precise that a student of bivalves, making allowance for known periods of rough weather, can count these deposits and estimate fairly accurately the age of a bivalve.

Shelter, of course, is a prime requirement of nearly all living things, and some insects have learned to construct portable homes. The larvae of certain leaf beetles use their own sticky feces to build or enlarge protective homes. The larva of one of these leaf beetles transforms into an adult still looking like a fecal pellet, and this deceptive appearance fools birds that would otherwise eat it.

Having the capacity to build, the shelter-makers do not have to do much to deceive. Some caterpillars are able to fool sharp-eyed chickadees by building houses in a multitude of different shapes. The cigar case-bearer produces an elongated shelter that is swollen in the middle and rounded at its closed top end. In the same tree, the pistol case-bearer makes a peculiar sanctuary that looks like a miniature gun. These strange objects appear dead to hunting birds which pass on in search of a meal. Bagworms make cases of silk strands and vegetation as soon as they are hatched. They drag their cases with them everywhere and keep building them up as they grow. After the males pupate in small shelters, they go flying in search of a female. The larger female, almost gross by this time, has no need to leave either the bag or her pupal skin. Instead, she breaks open one end of the bag, and her scent brings the male to her. Mated, she withdraws once more into her bag and lays a great mass of fertile eggs, though no movement is seen until the caterpillars hatch and leave the bag.

The attachment of the female bagworm to her home symbolizes the ability of nonmigrants to stand fast, using all their inherited capacities and adaptive abilities to survive when the outside world becomes hostile. To be able to withstand intense cold or heat is, finally, the supreme test of an animal's internal system; all the complex factors involved in its welfare must be working harmoniously if the creature is to meet— and conquer—such extremes of temperature. In their time of testing, the standfasts demonstrate the infinite flexibility and efficiency of their inner lives.

The Drive to Mate

Photographs by:
J. Behnke/ZEFA, 144
Martin M. Bruce/Tom Stack &
Associates, 145
Glenn D. Chambers, 143
Francisco Erize, 146-148, 155-157,
159-173, 183-187
H. Helbing/ZEFA, 174
Eric Hosking, 142
David Hughes, 141
Harold J. Pollock, 158
Edward S. Ross, 176
F. Sauer/ZEFA, 177
Robert S. Simmons, 178
Kojo Tanaka/Animals Animals, 149-154
Ron and Valerie Taylor, 175
P.H. Ward/Natural Science Photos,
179-182

The lives of all creatures are distilled into one great moment of supreme effort at the time of mating. Then, all the hereditary background of the creature's experience, its genes, and environment are tested in competition with the many other creatures mating, or attempting to mate. This is not always an idyll of procreation; it can be a bitter exchange, with some males eaten, stung to death, or badly injured. It is often a frenzied occasion when one dominant male drives off dozens of healthy, active competitors and claims ownership over all the females he can round up. It is the moment when a species matches the best males with the worthiest females. The Darwinian concept of the survival of the fittest seems to be given form during the act of mating, and we get new insights into the ingenuity, complexity, and mystery of animal life. It is a time when hormones begin to activate the different sexes at exactly the right moment, when chemical signals sent out into the ocean bring billions of creatures into breeding condition at a specific hour on a specific day. It is a time when the flashing lights of fireflies at the forest edge and those of creatures in the deepest parts of the sea help to match up mates and bring procreation to its fulfillment.

It is a time when complex safeguards work to prevent crossbreeding, overbreeding, or underbreeding. It is a time when lions fight, mountain sheep charge one another, antelope lock horns in bitter battles, and tigers savage other males.

But it is also a time of gift-giving. Penguins offer scraps of plants to females, and male flies proffer freshly caught insects to their prospective mates and thereby reduce the likelihood of being eaten themselves. In this time of bluff and of play, of singing and of dancing, of danger and of death, the sexes come together to create new life on earth.

141

The male Anolis lizard (141) vigorously pushes up its body and stretches its distinctively marked throat in a territorial display designed to gain the attention of a female and warn off male intruders. The edible frog of Europe (144) has a specially harsh voice, which is amplified by a pair of vocal sacs that bulge from each side of its head when they are fully inflated. The male toad (145) is less showy but just as effective in his efforts to attract a mate. He trills a distinctive cry in shallow water, and its vibrations send out concentric ripples.

The role of the voice in mating is so important that, when the male prairie chicken of North America (143) is seeking a mate, he appears to be disfigured by the resonating air sacs bulging from the sides of his neck. The peacock (141) prefers display when he wants to attract a mate.

Courtship Display

142

144

145

143

149

146

150

147

148

151

The classic confrontation preceding the mating act is between two males seeking to dominate a group of females. This antagonism is especially common among seals, nearly all of which must come ashore to breed. On land, competition among males for females and territory becomes fierce. Two giant male elephant seals (146-148) fight on a small section of a Mexican west coast beach. They use their six-thousand-pound bodies, blunt short teeth, and heavy snouts in a bitter contest for territory and dominance over as many as thirty females.

Fighting among lions is noisy and dangerous, and the battle may be terminated by the death of one of the animals when a nomadic lion attempts to gain entrance into a pride (149-151) where he would lead a more secure life. An old lion has little chance of holding his dominant position in the pride. Sooner or later he will be displaced by a younger, stronger animal and will be killed, or doomed to the life of a nomad for the rest of his days. Thomson's gazelles are aggressive creatures during the mating season (152-154), although they rarely hurt one another because their horns are not shaped to kill. The victor will mark the boundaries of his territory with a secretion from scent glands located in front of his eyes, as well as urine and feces.

In all penguin colonies, territory and dominance are important. Less dominant creatures tend to be pushed to the outer fringes of the congregation where their nests are more vulnerable to predatory birds. When two male Magellanic penguins (155-157) struggle for territory during the breeding season on a beach along the Straits of Magellan, the outcome of their confrontation will affect the survival chances of the loser's unborn youngsters.

152

155

156

153

157

154

Ritual is designed to bring two creatures closer together, to bond them more securely, and to harmonize their sexual readiness. In many instances, they also help the female to prepare the place where she will lay down her eggs. The male mallee fowl (158) takes the initiative by collecting an immense mass of grass and leaves and mixing it with earth to create a compost heap of rotting vegetation that will act as an incubator. But he must also attract a female to his carefully prepared mound, and he does this with low crooning sounds that encourage her to lay her eggs in the hole he has dug in the mound. With the eggs emplaced, he must work frantically to keep the incubator at the right temperature; he digs the mound out when it becomes too hot, and piles on more vegetation and earth when it cools. The waved albatrosses of the Galapagos Islands (159-163) must go through a formalized series

159

160

161

162

158

163

of posturings and gestures before they are both in the correct condition to mate. Both birds will collaborate in tending the single egg and the hatched chick until it is grown and independent. Mating behavior takes many forms, because some creatures need to be reassured that they are, in fact, dealing with members of the opposite sex. Others need to strengthen an already existent pair bond. Even when it is transitory, bonding is particularly important to birds, and the yellow-eyed penguins of New Zealand (164-167) have their own specialized rituals to synchronize their courtship movements and give them a growing familiarity with each other. Tiny details of plumage, of gesture, and of voice, become ingrained in their memories so that they will stay together and mate successfully. Many of the ceremonial acts are fairly obvious, such as the preening motions of nearly all birds when they are courting. The preen-

ing is a bond reinforcer. Many seabirds go through elaborate courtship flights, sometimes involving thousands of birds, the object being to bring them all into breeding condition at exactly the same time. This is to enable the birds, with their still flightless young, to desert their breeding islands in large groups over a short period of time. But some behavior is more difficult to understand. All gulls have ritualized courtships, but the male dolphin gull (168-171), with the courtship completed, screams continuously while he is copulating with the female. The white-breasted Peruvian cormorant is one of the famous guanay birds whose excrement is derived from the anchovy-rich waters of South America's western shores. The cormorant's courtship display (172) before his nest site not only appears to be an ecstatic tribute to his prospective mate (173) but is a vital part of courtship process.

164

165

166

167

168

169

170

171

172

173

175

174

176

177

178

The body structure of the mating creatures often dictates the duration of the mating time, and the act of copulation may last only a few seconds, or go on for hours. The readiness of the female to lay her eggs, or to receive a male, is also an important regulator of the sexual act. Sea snails which do not have shells combine both sexes in a single body. Two of these bisexual creatures (called nudibranchs) meet and lock together (175) while each supplies the other with sperm cells to fertilize their eggs.

No uniform rule applies to the act of copulation. A female skipper butterfly (176) spreads her wings and remains still, though facing in the opposite direction, while the male, with its more slender abdomen, transfers his sperm into her body. When ladybeetles mate (174), the smaller male must position himself above and behind the female, and the act of copulation is invariably greatly prolonged. This is also true of snout beetles, or weevils (177). Small frogs with poisonous skins, common in South American jungles, often remain clasped together for hours (178). The mating of praying mantises is a complex act. When the brown male mantis approaches a female to mate, he is already in jeopardy (179). The green female may seize him and bite him in half (180). She continues to eat him (181-182) until she has completely consumed him. Sometimes, a male mantis may successfully woo a female and actually begin to copulate, but the mounted female turns and eats his head and the front portion of his body. But the drive of the copulatory act is so great that despite being decapitated, the rear portion of the male continues the mating act and completes the fertilizing process. On other occasions, the female may wait until copulation has been completed before eating the male.

179

180

181

182

Mating with a Harem

Once the bull fur seal comes ashore and takes command of the territory on his island breeding grounds, he will not eat again until he has mated with all the females in his harem, a task that may take weeks. He claims his females as they come ashore pregnant and almost ready to give birth, waiting to mate with them until after they have whelped. The harem master moves in for a perfunctory courtship (183-184) and copulates with the first female which has given birth. His great weight and superior size make her helpless (185-186), but the moment she can free herself, she returns, pregnant again, to nurse her pup (187). Her uterus is forked and the pregnancy begins in the fork opposite the one that has just released the baby seal. After leaving her, the bull scans his harem once more. When the mating season is over, the exhausted and hungry bull returns to the sea.

183

184

185

186

187

As the buck impala assembles his harem on the high African grasslands, he must fight scores of battles with competing bachelor bucks, losing weight and condition in his effort to gather sometimes as many as forty does. But with his harem complete, even harder work begins. He must still keep the other bucks at bay as he leads his harem to safe grazing grounds, and he must respond to the mating demands of his does. At the height of the rutting season, his coat becomes ragged, and bones show through his once shiny skin. He looks bedraggled, almost defeated, but he must continue to fight and to mate in response to the ancient imperative that he impose his superior will over doe and competing buck alike. The buck impala is an extreme example of natural selection at work. The animal with the greatest strength and determination will perpetuate the genetic line. His predicament may seem comical to watching humans as he faces bachelor herds of impatient young bucks, anxious to express their own urge for dominance. It would appear to be much more efficient for the buck impala to have a smaller harem, with the does distributed among the other bucks, but the genetic heritage of the impala, stretching back countless millennia, has decreed that the buck must pursue his desperate course or be driven forever from the pleasures of harem ownership.

The simplest forms of life have no need to imitate the impala's frenzied behavior. The healthy one-celled protozoa reproduces itself by dividing in two, but these primitive organisms pay a price for their sexual serenity. Their evolutionary progress is limited to such mutations as may occur in what is, after all, a strictly insular world. Among the higher life forms creatures separate into males and females, with all the turmoil this divergence entails, giving natural selection a chance to work on a greatly expanded scale: With two parents involved each generation is a different genetic mix from the one before it and therefore has many more opportunities to improve the line. The quality and quantity of animal life is determined by the genes that are passed on. Since this is so, the search for a mate embodies all the vitality and mystery of life. Here is laid bare, often in one wild and brief encounter, the powerful drive to procreate. The mating act may be violent, pacific, lengthy, or short, but one similarity emerges. Paired individuals live, on the average, longer than unpaired creatures, one contributing to the welfare of the other. Indeed, the lion with his pride of hunting females is largely dependent on them to kill his food. He, in turn, preserves the integrity of the territory against wandering nomads or neighboring prides.

The sharing of resources, however selfishly it is done, may have devel-

The Drive to Mate

oped when sexual reproduction evolved. Among the simple forms of life the mating of microscopic cells often seems like some kind of last resort. It frequently occurs when neither cell has much reserve of energy to carry on alone. The cells combine, complete the sexual fusion, and from this mating may come a slight variation of cellular form which is better able to cope with a changing world.

The simple forms of life give us tantalizing clues about the complex conditions in which pairing becomes desirable or possible. Protozoans like the paramecium and the tetrahymena can only pair when food is scarce and they are not distracted by having something to eat. Besides, the temperature must be exactly right. The water in which they swim must be tranquil and suitably deep since the cells need both room to move and plenty of oxygen. In addition, the time of day is critical, and ideally neither of them should have been sexually active recently. The cells must be of unlike mating strains, which we can regard as being comparable to different sexes. But after exchanging genetic material, these tiny motes of life return to the primitive style of reproduction in which a single parent becomes two new individuals who share the newly acquired genetic characteristics of the parent's transfer.

We know about these precise conditions, but many other factors have not yet been satisfactorily explained. For instance, when the paramecium and the tetrahymena are mating, they release into the water chemicals that stimulate mating behavior among the same or related kinds of protozoa. So far, no scientist has been able to identify these extraordinary aphrodisiacs. Researchers have synthesized substitutes that can produce some of the same responses, but the recipe for the animals' own lure is still elusive.

The substance released by the nubile female codling moth consists mostly of unsaturated alcohol. When this chemical is manufactured by man it can attract male codling moths just as well as does the lure of the virgin females, but when the synthetic lure is used in competition with the real stuff, almost all male codling moths will try to reach the live females. Here is proof that in the natural lure is concealed some vital element missing from the synthetic substance.

Innumerable chemical lures are used by animals of the sea. All the oysters located along a four-hundred-mile stretch of shoreline, for instance, are immediately sensitive to signals received from their colleagues. The flavor of the water tells them when another oyster has begun to discharge sex cells. This trigger immediately synchronizes the physical states of all the other oysters it reaches and greatly increases the chances that traveling sperm will impregnate. Such signals, ranging just as far and with similar power, animate sea urchins, sea stars, and sea lilies. Sometimes man can exploit this process by taking suitable shellfish, breaking their bodies open, and throwing them overboard, thus stimulating spawning action along an entire coastline.

In Jamaican waters, and at various reefs around the Caribbean, tropical stinging sponges also need to detect the sperm being discharged by other colonies before they act themselves. Upon receiving the message, each sponge expels up to 20 per cent of its total bulk in a great cloud of reproductive cells. These cells, billowing like smoke, move in a massive chain reaction from one stinging sponge to another.

Closely related sponges are fertilized by this pervasive system of insemination. The sponges have an exact sense of the season, the time of day, and the stage of the tide. Their interior clocks are so precise that scientists can predict within an hour the moment when a sponge will begin its discharge of sex cells. Sponge colonies maintain this computerlike balance with no nervous system or body organ that, in man's estimation, could be capable of detecting, much less coordinating, such vital clues from the physical environment.

In the depths of the ocean, more complex creatures use touch and luminous markings to find their mates. Some underwater animals use electrical pulses to send sex signals; most whales, porpoises, and shrimp conduct two-way vocal conversations that help bring the sexes together.

Even the most cursory study of animals shows how complex and important a role the sex drive plays. While most sexual contact is transitory and casual, and lacks the feeling human beings associate with sex, some elephants are an exception. Frequently two animals seem to form an attachment even before the urge to mate becomes imperative. When the "special" cow is ready to copulate, she arouses the bull with provocative body movements, then repulses him when the impassioned animal attempts to mount her. This encouragement-and-rejection performance goes on until the final act of the elephants' erotic foreplay, when the cow uses her trunk to caress the bull before they mate. Other animals also have long-lasting sexual relationships, and an ancient subject for discussion among behaviorists is the value of these enduring sexual bonds.

In human cultures where monogamy is traditional, the rationale is that it is better for the children to have willing men and women democratically divide responsibilities to perpetuate the society. But polygamy has an even older human history. In polygamous cultures it is incomprehensible that every man should aspire to have a wife and children. Many men, it is pointed out logically, are not able to support dependents, while others can provide for several wives and their children.

But comparisons between men and animals tend to be invalid because of the difference in the length of time that a youngster is cared for in human and nonhuman groups. There is no comparable dependency period anywhere among animals, so perhaps it is significant that non-human life finds fewer rewards in long-lasting bonds between male and female. We know of only a few instances in which the pair bond is maintained as long as both partners remain vigorous. Year after year, the beavers occupying a lodge consist of the same mother and father and their young. Year after year, the same dominant male wolf and his dominant mate lead the pack, which is well served by having the most vigorous wolves as the leaders and the main procreators. They are followed by their young and any newcomers that manage to join the pack without upsetting the social hierarchy. The pair bond continues until one of the dominant leaders is injured or weakened by disease or age.

The reason for fidelity among other animals is less easy to understand. It is practiced by such disparate groups that no common denominator can be divined. Why should jackals and moles share the attribute of fidelity with marmosets and gibbons, some whales, most pigeons and parrots, geese and swans, ravens and some tropical birds? Then there is the added mystery of fidelity among fishes like the cichlids and the butterfly fish.

There is an extraordinary similarity in the vocal repertoires of monogamous tropical brids. All fifteen species of African shrikes, along with a scattering of unrelated birds in other parts of the world, are duet singers. One of the best known is the bell shrike of East Africa, which has been intensively studied by W. H. Thorpe, of Cambridge University. He has found that one bird of a pair, usually the female, begins the song. Then she pauses while her mate utters a note, or continues to sing while he joins in with almost perfect timing, either at the same pitch or at another one that produces a musical chord. A musician might call this performance singing in unison, polyphonal activity, or antiphonal dueting. But a pair of bell shrikes are capable of all three acts. Each singer seems alert to every sound from the other bird, responding in tempo, even though they cannot see each other. Thorpe feels that the singing keeps the pair in constant touch while it simultaneously stimulates the pair bond.

Thorpe laboratory-tested a prediction made by Konrad Lorenz that if one member of a dueting pair of birds were to disappear, the other would complete the song, singing both the notes of the missing mate and its own. In Thorpe's experiment, the "deserted" bird did indeed complete the singing, as if it were being performed by both birds. But after a week alone, the performance began to deteriorate. Then, if Thorpe played a recording of the bird's own song, the only response was a brief repetition. But a single note played in the voice of the missing bird was enough to restore the high quality of the earlier singing. The listening bird could identify its missing mate's voice in half a second, an ability it shares with many birds that nest in groups where individual recognition may depend on exceedingly brief and subtle signals. It is not known whether any mammal uses so short a sound to any advantage, but we must assume that it is possible.

Polygamous creatures do not need instant identification, since most animals have only brief liaisons and are generally promiscuous. Until a few years ago, apiculturists thought they could improve the honeybee lines by controlled breeding, but recent research has shown that the virgin queen on her mating flight has catholic sexual tastes. The first drone to catch her gives her only about one-sixth of the number of sperm cells she will need for her life of egg-laying. He pulls away from her, tears off the reproductive parts of his body, and dies soon after he hits the ground. The flying queen expels the remains of the drone and accepts a second lover. She repeats this performance until her storage sac is filled. There, the sperm from six or more successful males is thoroughly mixed, and chance alone determines which male fathers any queen of the next generation.

The ordinary housefly and the yellow-fever mosquito, on the other hand, receive sperm cells from only one male. The buzzing sound of the mosquito's flight attracts every male within hearing distance, and each male able to grab her would copulate if she would let him. But for the first forty hours after the female mosquito emerges from her floating pupal case, the concentration of juvenile hormones in her blood is still so high that it inhibits the nerve centers that control her mating responses. A dozen different males may attempt to inseminate her, but she won't cooperate. When she finally does respond, the deed is done quickly, and at the moment of insemination she receives a substance from accessory glands in the male that puts an end to her willingness to accept any more sperm. The substance, which has been named *matrone,* is a chemical chastity belt which makes the female permanently unreceptive, though, oddly enough, it does not end her attraction for males. They are still aroused by the sound of her wings and rush toward her. The housefly, which receives the same accessory substance, no longer releases her seductive fragrance, and males ignore her.

Many problems are resolved during the courtship of animals. Perhaps the male spider developed his complex courting sequence mainly to persuade the female not to eat him. The nervous tension that builds up in drakes before they conquer their ducks results in furious preening, though each species may have a different preening gesture. The signals of intent must be precise and they must be understood by the other creature involved, because the difference between gestures of attack and courtship are sometimes difficult to interpret, and females often need reassurance about the intentions of the approaching male (though it may be difficult for man to understand what is reassuring about the noisy courtship of humpback whales that swim together, giving each other tremendous slaps with their powerful flippers and sometimes hurling themselves out of the water as they prepare to mate). In gentle contrast, Falkland sea lions come ashore and use their undulating necks to fondle and touch each other. Occasionally their mouths meet in what can only be described as a kiss. Female gulls toss their heads to incite males to mount them; male sticklebacks prod females to make them spawn. The croaking of frogs brings males and females together, though the male may not be sure which is a female and which is a male. Only when he grasps another male does he learn his mistake. The gripped male grunts, and the sound is enough to dislodge the amorous hold of the other frog.

Merely the finding of a mate of the right species may demand special sensing equipment. The pine chafer beetle has a pair of olfactory receptors that look like miniature flags sticking from its head. Apparently these are used to locate and court females. In the dark of a summer night, large-eyed male fireflies zigzag back and forth above a field, intermittently flashing their abdominal lamps. Their nervous systems control the pulses of light that produce a luminous wink, a streak, or a modulated message, depending on the species. A watching female waits on the tip of a branch or on a blade of grass. If she is ready to mate, she signals back to the flashing male. This brief wink of light not only reveals her position and her availability but also identifies her species. If her luminous signal comes a trifle too early or too late, in relation to the male's flash, he knows she is of the wrong species, and ignores her. But if her reaction is pitched to exactly the correct number of milliseconds, the male turns and flashes his lamp again. A few interchanges of light lead him in for a landing close to the female. Then odor and touch replace the light signals, and mating begins.

Although this signal system is extremely accurate, it is not always perfect. The female may attract a male of the wrong species by being a millisecond off in her signaling. If this happens, she eats him. Some females put out a series of wrong signals, and many unlucky males that do not give off the right stimulus go to their deaths, while the females' eggs remain unfertilized. The imagination of the observing man is gripped more by visual signs than by scent lures because humans are so visually oriented in sexual behavior. Man is better able to appreciate the light of a glowworm than the seductive and aromatic messages of a moth. When the female glowworm is ready to mate, she curves the underside of her body, like a scorpion raising its tail, and reveals fluorescent light panels. Males flying reconnaissance six feet overhead see the signal that she is ready to mate, but not all respond. We do not know why one light should be more attractive than another, but we do know that a smitten male drops like a dive bomber. The light of the female becomes his target and he crashes into her noisily, if his aim is good. The collision between the two insects may send the female sprawling, but mating begins as soon as she rights herself. The mission accomplished, the female glowworm's light goes out, her need to signal fulfilled.

The mating of snails gives us a fascinating insight into the subtlety of signals and responses. Snails are hermaphrodites. Yet the fact that snails possess the organs of both sexes does not fully explain their peculiar mating ritual. Two snails circle each other for an hour or more as a prelude, apparently, to the firing of what may be called love darts. Both snails produce these tiny, sharp darts, which are about a centimeter long and are made of a carbonate of lime. The dart is fired by a propellant that is stored in a glandular chamber just behind the creature's eyes; the snail makes a slight hissing sound as it ejects the dart. Usually the dart is not harmful, penetrating only about one millimeter into the other snail, but if it accidentally strikes a soft part, it can reach vital organs. The love darts apparently have something to do with synchronizing the mating process. The first creature to fire the dart becomes temporarily passive, while the other snail continues to show its desire by circling before it, too, ejects a dart. This strange foreplay is mutually exciting and eventually leads to copulation that is just as odd. Each hermaphroditic creature introduces its male sex organ into the female orifice of the other.

Love darts are bizarre enough, but the mating habits of the octopus are scarcely more credible. In his *Historia Animalium,* Aristotle wrote:

"Some assert that the male has a kind of penis in one of his tentacles, the one in which are the largest suckers; and they further assert that the organ is tendinous in character, growing attached right up to the middle of the tentacle, and that the latter enables it to enter the nostril or funnel of the female." When a male octopus approached another octopus, he added, the creature would exhibit oversized suckers and display his sex in colorful stripes and spots on his gray body. If the second octopus were also a male, he responded in the same way, but a female would be more passive, perhaps drawing her eight arms up close to her body and then pushing suddenly in an underwater, eight-point kick that violently rejected the approaching male.

Martin J. Wells and his wife, zoologists from Cambridge University, experimented with a male and female octopus separated from each other by a clear plastic partition in a saltwater aquarium. The two animals explored their limited quarters visually and by touch. Through a small hole in the partition, the two creatures discovered each other and when the female did not move away, one of the male's arms explored her body. He became sexually excited and went into his full display. He reached with his third right arm into an opening in his body and picked up a packet of sperm. He pushed this through the hole and put it into the mantle cavity of the female, releasing it close to the opening of her egg duct. He had found another octopus, and learning that she was a receptive female, he had given her all she needed to fertilize her eggs. Off the coast of Maine on Matinicus Rock, an elliptical sanctuary with a great assembly of sea birds, the possession of a nest site gives the male birds their main drive to go ahead with mating. Without land, the bond between the paired sea birds would be useless. During the premating period, males advertise their presence with distinctive species calls which tell the females that they are available, that they own important territory, that they are ready to fight trespassers, and that they will cooperate with any female, provided she is sufficiently submissive. As soon as he sees submissive behavior, he knows he is dealing with a female and modifies his aggressive stance as the first step toward mating.

The territorial songs of birds are individual, at least to the informed human listener. But the massed cries of toads and frogs in a spring pond, seemingly a cacophony, are just as specific as bird calls and signal every member of the species to an orgy of clasping, egg-laying, and fertilizing.

Sound is only one of many mating signals. When smooth-skinned salamanders slide over dew-wet ground and slip into fresh water, they are being led on just as surely as the sea birds that follow calls rising above the sound of rushing waves, or frogs responding to urgent croaks from a small pond. But the salamanders are apparently guided by scent, and they usually return to the same waters they left to begin their dry-land lives. Once in the water, the males prowl the bottom and deposit small pyramids of sperm-filled mucus. The females pick up these offerings with sensitive cloacal lips and take in the sperm, fertilizing their eggs. The silent underwater ballet is soon over; the eggs laid by the females wait to hatch while the parents leave the water to resume their dry-land lives.

Successful pairing, however, depends on more than mere identification and proper signaling. The timing of the mating act must be coordinated between the sexes. The wrasse, a three-inch fish found off the coastal reefs of eastern Australia, was for many years a biological oddity because it spent so much of its time removing parasites from the skin of larger fish. Indeed, it was so much in demand by the fish that they literally lined up for its services. But recently it has been discovered that the male wrasse defends a definite territory in the coastal reef and holds dominion over several sedentary females, sharing the center of the territory with the largest of them. His other females have their own hierarchy, depending on their size, but if the male dies and there is no other male around to seize his territory and his harem, the large dominant female begins to take on the characteristics of the dead male and in fewer than twenty days she has taken over his job. Dormant tissue in her ovaries develops into testes which produce sperm. Her ovarian tissues shrink and become useless. As her transformation into a male wrasse is completed, her position in the harem is taken over by the next largest female. Eventually, the new harem mistress may inherit the place of the transformed female, going through the same biological changes to keep the species intact and to preserve its strange behavior. Constant sexual vigilance is necessary to prevent crossbreeding, and a kind of fail-safe system cuts the chance of such accidents to a minimum. Apart from specific gestures, sounds, and smells, for instance, the genital organs of many insects and crustaceans are so specifically detailed that scientists use them to identify species. The fail-safe system works by ensuring that the male is exactly the right size and shape to fit the female, which is why the genital organs of animals conform much more in size than do the other parts of their bodies.

Male reptiles, with the exception of the unique tuatara, keep their sex organs hidden until they are ready to mate. The organs are secreted in the base of their tails like the inverted fingers of a glove. When the erect penis is pushed out through the cloacal opening, the inner wall of the penis becomes the outer wall. (The tuatara, once mistakenly thought to be a large-headed lizard, is the sole survivor of its ancient kind, living now on a few New Zealand islands. It lacks any penis at all and mates by bringing genital openings together, suggesting that millions of years ago reptiles mated in this way.) Male lizards and snakes have two hollow sex organs in their tails. Though both of these penises are operative, the creature uses the one which is closest to the female's orifice. When this organ is extended, a furrow channels the sperm into the female, and the surface of the erect penis may be creased or spiny to help it remain in place while the process of fertilization is going on. Male sea turtles have a difficult time just getting mounted on their slippery mates, and many clumsy attempts are made before success is achieved. But once on top, the sea turtle makes sure of his position by gripping the front rim of her shell with his two big claws and curling the tip of his strong tail under the back part of her shell.

The two-inch-long earthworm has a much easier time of it. Extending its body from its burrow opening, it feels in the darkness for another worm of the same kind and size. It has no need to find a mate of the opposite sex, since each earthworm combines in its body the organs of both male and female; but it must find one that is the right size. If its nearest potential mate is eight inches long, mating is impossible, and the small worm must search until it finds a creature its own length. Consequently, much of the earthworm's mating life is spent looking for the right sized partner. Snails and slugs abide by the same rule and gain nothing by trying to mate with creatures that are either too large or too small. The largest snails, rare in any population, compensate for their infrequent mating by producing greater numbers of fertilized eggs. In tidal pools and freshwater shallows, tiny, curved-bodied scuds adhere to a different pattern of behavior. The male scud is invariably larger than the female, and he carries her on his body until she is ready to lay her eggs. Unmated males grab any individual smaller than they are, but if the captive struggles to escape it is clearly a male and is released. The more passive female curls up, making it easier for the male to grasp and carry her. But he must find a partner that is almost exactly seven-tenths his own length. He never tries to grab one that is larger, or one that is too small.

While some animals are unable to make immediate sexual distinctions, birds have no such identification problems unless man interferes. The male yellow-shafted flicker will court any female that approaches him, unless a scientist has added a black mustache to her face. Despite her submission, the male will attack the mustachioed female, and will kill her if she doesn't flee. When the bright red epaulets of the male red-winged blackbird are covered with black paint, the bird can no longer bluff competing males into respecting his mating and nesting claims. Even his own mate may treat him as an outsider.

The power of scent is everywhere in evidence. The male monarch butterfly lands on a leaf and displays two tiny black scent sacs swelling on each of his wings. He has already lured a female to him by releasing another scent from special glands at the tip of his abdomen. She has followed this scent and has settled close behind him. Now he surrounds her with scent to stimulate her sexually, and he quickly daubs parts of himself with the scent from the sacs on his wings. These secondary scent-makers hold the attention of his mate while he extends a pair of forcepslike claspers which seize the tip of her abdomen. In a moment the two creatures are copulating. The scent sacs on his wings were used only briefly, but they were vital to his mission.

The antennae of certain male moths hold sensitive olfactory organs which can detect the smell of a potential mate even though she may be miles away. Males of some species can find a female even in a town burdened with a bewildering variety of smells. The male manages this upwind mating flight by discriminating between the number of her scent molecules reaching each of his feathery antennae. If he loses one of these feelers his directional capacity is killed. He can still smell a female, but he can no longer find her. The female moth also needs an olfactory sense to check the sex of her suitor and to identify the plant on which she will lay her eggs. The moment she is mated, she stops releasing her alluring scent and begins the nightly deposit of large numbers of eggs. Her behavior changes because she has received, along with the sperm, a chemical substance which triggers the production of a hormone in her body. It is this hormone, rather than the physical act of mating, that alters her behavior.

Compared with the sensors of these male moths, the antennae of honeybees look much simpler. Yet they are deceptive, too. Honeybees can measure an immense number of different fragrances, and their antennae help give them the capacity to distinguish between drones, workers, and virgin and mated queens. The honeybee virtually lives by scent. The queen bee's abdomen contains glands which release fragrances that influence hive members. Scent stimulates workers to cooperate to preserve the colony. Drones contribute their own scent, which probably influences the workers to build special cells in which new queens will develop. The drones' scent stimulates the old queen to depart with her swarm of followers, and when the virgin queen makes her round-trip flight from the doorway of the hive a few days later, she is followed by a trailing tail of drones, all of them encouraged by the powerful scent she offers.

Scent and markings guide a mating pair accurately enough to their vital meeting, but neither device is of any use unless both partners are in breeding condition. And to reach such a state demands a complicated series of physical reactions inside the body of each. These reactions may be triggered by internal clocks or by signals from the outside world, or by both.

In the republic of Panama, most local birds do not begin courtship until the millions of winter visitors have left for spring breeding in the north. As long as the Panamanian birds must share food supplies with so many migrants, they feel no urge to nest. But as soon as they have the local resources to themselves, they find mates, court, nest, and raise new generations. Yet the environment they are living in has no measurable seasons. Instead, they inhibit their sexuality until their signals tell them there will be enough food for all.

Creatures of the sea respond quickly and prolifically to cyclic abundances of food. Some algae-eating abalones and chitons living in the waters of Puget Sound accumulate a stockpile of nutrients in their storage organs while food is abundant. At least two kinds of predatory sea stars, which attack barnacles, limpets, and mussels, do the same thing. When they are ready to make sex cells they transfer the nourishment to reproductive tissues and change their behavior.

Both sea worms and fish appear and mate on regular schedules that may be influenced by season, phases of the moon, stages of the tide, and time of night. Certain paddle-footed sea worms are transformed; their eyes become larger, and extra stiffening bristles sprout in their paddles. They stop hunting and scavenging for meat. Their digestive organs shrivel as the maturing eggs of the female and the sperm cells of the male fill the space inside the body wall. Then comes the synchronizing signal that animates the worms; the phase of the moon tells the worms when the time is right. They "know" that at the full moon, or the new moon, the tides will be at their greatest extremes. Most of the worms make no mistake, though some may be one night early, or one night late.

The worms leave their burrows in the mud at the bottom and head for the surface. The smaller males travel fastest and arrive first. They curve back and forth through the water exploring the upper reaches of this new world while the larger, stiffer females drive themselves obliquely upward along straight paths. The males sense their coming, and the curving movements increase. Each male swims in a spiral path around an approaching female's body. The long bristles on his paddles brush against her bulging skin. She explodes, expelling eggs through rips in her body; the eggs join a flood of milt the male pours forth in response. Both creatures quite suddenly become almost empty bags, which settle slowly to the bottom and die. Those fertilized eggs that survive the predations of countless hungry hunters then develop into a new generation of paddle-footed worms.

The palolo worms living around the Dry Tortugas off the coast of Florida and among South Pacific atolls are so precise in their mating schedules that local people who relish the females can plan their festivals to take advantage of this special delicacy. In October and November, at the third quarter of the moon, there are palolo risings, which feasting South Sea islanders attend. When scientists studied the palolos, they discovered that each worm rising to the surface was headless. It paddled in reverse and arrived tail end first, leaving its front end concealed among the corals on the bottom, where it remained an active predator and scavenger. Only the last two-thirds of the worm's body had been transformed in readiness for the sexual encounter, but this tail end, now independent, had developed a pair of eyes surprisingly sensitive to light. The paddles on the swimming portion of the worm had enlarged as eggs and sperm matured. Then, on the night of the rising, the worms divided themselves unequally, sending the reproductive parts of their bodies upward to perform their mating water ballet just below the dark surface of the sea.

Smeltlike grunions come out of the waves of the Pacific coast at night from Monterey, California, southward, ready to follow their ancient tradition. They appear when the tide approaches its highest point, though each grunion makes only one trip a year. The swarming rituals occur every fourteen days from April to August and the grunion schools wait offshore for these crucial moments. When the time comes, they allow themselves to be cast on the wet sand at the edge of the advancing waves. Each female wiggles tail-first into the sand, almost burying her four-inch body. Instantly, as many as three male grunions flood her eggs with milt. The next wave carries the scurrying fish back into the ocean. For the next two weeks the eggs lie in the sand, untouched by waves, unless there is a great storm. They are ready to hatch within a week but they are programmed to wait until the high night tide reaches up to them and water seeps through the sand. At this signal, the waiting embryos pop out of their shells and are washed back into the sea, to become adult grunions, if they are not eaten.

One visible part of the network of signals that govern life was witnessed by Christopher Columbus. On the night of October 11, 1492, at ten o'clock, crewmen on the *Santa Maria* sighted a peculiar light in the sea. The moon was within a day of reaching its third quarter, the time when luminous worms of the West Indies swarm at the surface of the sea. The light spread across the water as the female worms extruded both eggs and a luminous secretion to attract and stimulate the males, which rushed forward, flashing their own light, and fertilized the eggs.

Courtship gifts and the sharing of food are common ways in which animals reinforce the bond between mates. To his intended partner the male Adélie penguin brings a stone, his first contribution to the crude nest that will be built on the cold ground of Antarctica. The male of one parrot family moves his head convulsively and regurgitates seeds, which he offers to his female. The dog fox delicately approaches a vixen and drops a mouse at her feet. The reward in all such interactions is essentially the same, but the variations on the theme seem endless.

A Harvard University entomologist, William Morton Wheeler, traced the history and type of gift exchanged between male and female dance flies. The tiny creatures he studied showed the evolutionary process in dramatic detail. Few of these insects have a wingspan greater than half an inch, but they perform ritual flights in large social groups. During these flights, males identify potential partners and follow them to resting places. While the two flies stand on a leaf, the male of one species offers the female a freshly caught smaller insect. With her attention focused on this substitute, she is less likely to eat the male. As she sucks the gift dry, he copulates with her.

The male of another species adds frills to the gift offering. He prepares his present by swathing it in silk fibers from his salivary glands—and gains himself extra mating time while the female tears the gift apart. The male of a third species fashions an even larger salivary silk package, but this one is all show; it contains no food at all. He offers his pacifier to the female, which toys with it or unravels it layer by layer. Frequently, she is still busy with her token gift long after he has finished mating and departed.

For creatures that mate at any time of the year there are usually cyclical peaks caused by social stimulus. George Schaller's lion studies reveal, for instance, that when one lioness is ready to mate, other lionesses in the pride are often stimulated by her condition and are quickly ready to mate. When new males take over a group of langurs in India, many of the female monkeys become ready to mate simultaneously.

Grazing animals must have an abundance of food before mating. When drought strikes the great grazing plains of Africa, many hundreds of thousands of female antelope never come into breeding condition at all. Abundance conditions the animal and stimulates its glands to action. The bull elk becomes fat in readiness for the rut, but by the time it is over, he has used up all the fat. By contrast, his placid cows remain plump, for they have been putting on weight since midsummer, when they weaned their calves of the year before. Their time of testing comes in the spring, when they are nursing new calves, but food is then usually abundant.

During the rutting season, a cow elk is stimulated by the bugle calls of the bulls, the snorting of young males as they clash antlers in shoving contests, the scent of bull-elk urine (and perhaps hormones in it), and the sight of a dominant animal posing conspicuously to display his size and antlers. The cow elk may dodge a bull until her time for mating comes. She rarely runs far, but she may kick the male if he presses too close. At times, she responds to his advances by stopping, moving her hind feet apart, arching her back, and discharging her urine. The bull often thrusts his nose into the stream to test its flavor to learn how soon the cow will be ready to mate. He then raises his wet muzzle and curls his upper lip in a gesture common to most males that grow antlers or horns.

The mating of grazing animals is as diverse as the pairing of other creatures. Wildebeest roaming the plains and savannah of Africa face great difficulties before they can copulate. Bulls may have to establish territory and hold it against other males. They must preserve harems of cows and calves. They may have to migrate great distances to find new fodder, and they may be harried by lions, wild dogs, and hyenas.

Another antelope, the Uganda kob, has an easier time of it. The kobs annually interrupt their sociable roaming of open savannah country to establish mating stations, called leks. Here, males establish a hierarchy with the dominant male at the center. Mature females wander into these designated territories as hormones bring them into season. Most of them refuse to mate with any except the most dominant male, but when he tires and leaves his favored position, his place is taken by the second most vigorous male. Although the element of choice is restricted, the system ensures the simultaneous readiness of both partners in every encounter. A rowdy group of hunting dogs pursues any bitch whose odorous lure attracts them. They bluff or fight to establish their own hierarchy, which is complicated by the fact that the dominant animal is almost always a female. Each time the dog approaches a bitch in heat, she turns sideways, thus retaining the option of using her teeth if he presses too hard. This stratagem is a play for time, so that the ovaries can complete their production of eggs and the uterus can ready itself for the growth of embryos. Hormones from these organs presumably operate through sensory centers in the brain to adjust the behavior of the female as she develops her ability to conceive.

The sex act is not the end of the sexual process among most animals. Many females need some rest while the sperm make their way to the eggs. Among some creatures, including some rats and a number of insects, part of the semen injection serves to produce a vaginal plug, which seals the sperm inside and may interfere with copulation for a while.

Many males need rest too. The cricket becomes silent and will not resume calling until he regains his ability to copulate once more. Then his cries rise in a crescendo that is an exact gauge of his capacity to mate. By contrast, the male rat sings a peculiar post-ejaculation song which a female rat can hear but which human beings cannot without the aid of instruments. As he recovers from the sex act, his song ceases, and this is a sign that he is ready to mate again.

One of the more astonishing post-copulation acts is performed by mice. If a female mouse smells a strange male at close range during the first few days after her mating, she will almost instantly abort her litter. The miscarriage is caused by some unidentified substance in the urine of the strange male. In fact, if a small amount of his urine is brushed on the nose of a newly pregnant female, she will lose her young. The male mouse reacts quite differently to the presence of a strange female. She excites him, and he is quickly ready for copulation while his interest in his familiar mate wanes progressively.

Each step in the reproductive process reveals the complexity of animal behavior and shows us that nothing is quite what it seems. For every action there is counteraction, and for every signal there is a sensor to receive and interpret it. Sights, sounds, and smells all combine in an intricate interplay of courtship and mating that can only inspire the admiration of man.

Problems of Parents

Animal behavior is often revealed to human watchers with special insight when the animals themselves are absorbed with their most critical activity—rearing their youngsters. Then their efforts, similar to man's, can be understood more clearly. But the human observer cannot always draw facile conclusions from what he sees. The buck gazelle dashes forward when a jackal menaces a gazelle fawn hidden in the grass. It appears that the buck is trying to save the fawn. In fact, the gazelle is only trying to preserve the integrity of its territory. The buck has no mechanism, no concern, to deal with the problem of the fawn. It will likely crop grass nearby while the fawn is being eaten.

The drive to bear and protect young is genetically programmed and it conforms to the rules of inheritance. This can result in behavior that is both identical to human action, and completely unlike it. Some species of hummingbirds compete so vigorously for scarce nesting materials that they rip each other's nests to pieces and ruin the efforts of any of them to breed. Some parents simply abandon their youngsters, or their nests, at the first sign of danger. Others defend their youngsters to the death. Whatever behavior best suits the survival of the species occurs.

Animal parental care shows the great importance of the mother. She can not only rear, but also transfer her experience when the youngster learns to imitate her. And parental care is not confined to the higher animals. Some of the fiercest defenders of their young are fish, spiders, and other non-mammals.

188

189

The Egg Sac

The egg cocoon is the spider's most effective tool in preserving the species. This spider (188-194) is seen in the process of constructing an almost perfectly spherical egg cocoon hung from a twig. The spider had spun a flat web blanket, laid her eggs on it, then bound the blanket together with another sheath of weaving.

Inside the cocoon, the spiderlings will hatch with their mid-guts filled with yolk, which will act as food. They cannot eat as long as this yolk is undigested. Once this digestion is complete, they will mature, shed their skins, and be ready to begin life as hunting animals. Then, the spiderlings must escape from their case or immediately attack and eat each other.

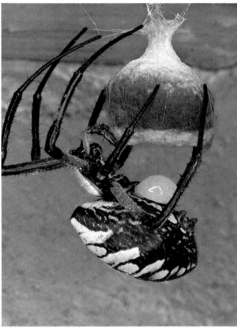

190

191

192

193

194

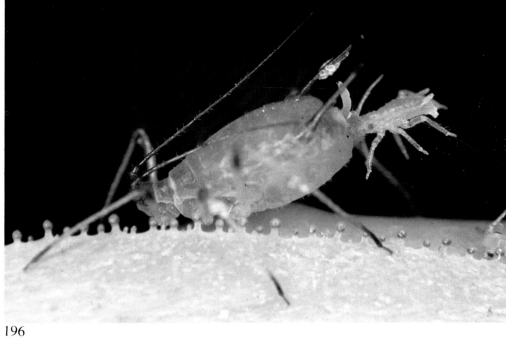
196

195

197

To provide shelter for the eggs and youngsters born to the breeding creatures is a test of ingenuity. There is no uniformity of treatment. Some young creatures are totally protected, like the spiders; others are cast free and loose into the world to fend for themselves as soon as they hatch. The birth of the aphid (196) sees the young aphid emerging live from its mother's body, standing on the stem of the plant from which the young aphids will immediately begin sucking juices.

Many wasps build mud houses or shelters to protect their young (195) and this mud wasp has caught and paralyzed a boll-worm which will become food for the mud wasp grub. Butterflies use a variety of devices to conceal their eggs, including waterproof cases, but this creature, Araschnia levana (197), is using its delicate ovipositor to insert its eggs into the stem of a leaf. One device is to enclose the eggs in a thick gelatinous covering, or to fasten them so securely to underwater positions that they are not attractive to egg hunters. The Everglades snail (198) extrudes its large eggs onto the stem of an underwater plant. The quality of parental concern is not necessarily linked with evolutionary development. The python guards her eggs fiercely (202), so that when they begin hatching (203) they will be protected against enemies. Even among fish, most of which shed their eggs into the sea to join the floating plankton, there are individuals which guard their eggs, as demonstrated by this carp (199). Some even build underwater nests or shelters to guard their eggs and young. The eggs of stink bugs are contained in beautifully designed canisters and the stink bug guards the eggs (200) until they hatch. The nudibranch, or sea slug (201), encases its eggs in a long ribbon.

202

198

203

200

199

201

204

205

In returning to the sea, after spending a long prehistoric transition period in estuaries close to shore, the dolphins became true ocean-going mammals. Their intelligence, their capacity to cooperate, their ability to communicate, are all a part of typical mammal behavior, exemplified here by the underwater birth of a dolphin. The female remains motionless while the tail of her youngster appears first (204). Then, waving back and forth, the youngster gradually appears (205).

The dolphin mother is not only devoted, (206) with her youngsters swimming close beside her after birth, but other members of the dolphin group will cooperate to ensure the safety of newborn young. A second adult appears (207) and flanks the newborn creature as added protection until he is strong enough and fast enough to hunt independently for himself.

206

207

Mammals and Young

Mammal mothers watch closely over their young for weeks, months, or even years after their birth. Just as a house cat will move her kittens if she is disturbed, so may other mammals. African hunting dogs move all the young of their packs if their hunting territories change. The young elephant remains dependent on its mother for more than two years after birth. A European dormouse (208) carries her youngsters, one by one, to a new and safer nest. A young three-toed sloth (209) clings to its mother's stomach as she moves through the trees of South American rain forests. In Europe, a ratlike marsupial (210) carries its young on its back in the manner of an American opossum. Spider monkeys (211) hang by their tails, which they can use as a fifth limb, in South American forests. Victims of the prowling jaguar, they rarely descend from their treetops, and the youngsters are carried from branch to branch

by their mothers. Japanese snow monkeys—or macaques—beat the problem of often intense cold in their mountain habitats by staying close to a region of steaming pools, particularly when they have young to care for. They take their young with them for long warm baths during the coldest weather and so both protect the youngsters from the cold and teach them how to survive into adulthood (212-214) over a long nursing period. The North American opossum, like all marsupials, gives birth to almost embryonic young which crawl through her fur to reach the safety of her pouch, where she has 12 or 13 teats to suckle them (215). The young must return to the pouch and the teats to suckle even when they are old enough to ride on her back (216).

208

209

210

211

212

213

215

216

214

119

The Newborn Seal

The birth process for most mammals combines an instinctual physical response and almost human coddling on the mother's part. In a colony of South American sea lions on the coast of Argentina, a mother will protect her youngster fiercely (217-220). The birth is painful (217), and while the mother rests, the pup nurses (220). When a curious female approaches the pup, the mother rears up in a threatening posture (218). Her defense occurs close to the moment of birth, and the remains of the birth are still visible in the center of the photograph. Later, when the sea lion pups are older (219), they will learn adult behavior from playing together. They will imitate the struggles for dominance among the adults. The northern tundra is often covered with fresh snow after the caribou have left their winter shelters, and snow may fall well into the spring birthing period (221). A mother that

gives birth during the northbound migration can give her youngster no shelter against the wind. She must help it to its feet, let it nurse for its first meal, and then encourage it to keep up with the moving herd. Sickly calves get left behind, to fall victim to wolves, which usually follow the migrating animals. A female kangaroo (222-223) can nurse a newborn youngster in her pouch as well as the older "foot" kangaroo, which must follow his mother until he is weaned. Each uses a separate teat, the pouch youngster getting the richer milk. In times of severe drought, both pouch animal and the foot youngster die, but the doe kangaroo maintains a dormant blastocyst inside her body, which remains quiescent until the drought ends. Then, with feed available, the blastocyst transforms into an embryonic kangaroo and is born quickly. This saves valuable time.

217

218

219

220

221

222

223

226

227

228

229

230

224

225

For birds, the nest that is the focal point of life during egg-laying, incubation, and the raising of the young, only serves so long as the young need it. Then parents and young will separate. Both will travel widely before the next breeding season. Some birds return every year to the same or a nearby nesting place; others return to the same general territories. The redfooted booby (224) ranges widely over tropical oceans, diving for fish and squid, and then attempts to claim its own birth site to mate and make its nest. The great blue heron of the Galapagos Islands (225) puts on a splendid show of tail feathers in an aggressive display, defending its eggs against intruders of the same species. The common coot of Europe and western Asia does not need the help of its parents to tap its way out of its shell (226-230). When free of its imprisonment, it is dependent for weeks on the food brought to it by its parents.

The rufous fantail of Australia (231) feeds its young on fresh insects and continuously displays the feathers for which it is named. Once out of the nest, the young fantails soon are able to follow their parents' darting, twisting flight, and so learn to catch insects themselves. A young ground plover stands beside its parents (232), but, unlike the fantails, is not fed by its parents at all. It must go hungry until it learns by observation and imitation to find its own food on the ground. Flamingoes (233) breed in a mud nest shaped like a miniature volcano. A single egg is laid in the crater, which is just big enough for the brooding of one chick. One parent is always with the chick, instantly ready to regurgitate food for the hungry youngster. The different methods used by birds to rear their young are insights into the almost endless capacity of evolution to find different ways of perpetuating the species.

231

232

233

Egg Predators

Many sea islands birds cluster together, competing for a limited number of nest sites. Until the eggs are hatched and the young grow large enough to leave the nest site and group themselves into "creches," the parents must protect their offspring. Hunters harass them. Kelp gulls—the black-backed gulls of the southern hemisphere—may nest close to the jackass penguins of southernmost Africa and steal their eggs or even chicks to feed their own young (234-236). After the gull has threatened the striped penguin, the gull walks off with the egg in its mouth while the penguin searches its nest. The agitation of the parent penguin contrasts with the apparent unconcern of an albatross caught in the same dilemma on Hood Island in the Galapagos group (237). The albatross remains incubating while three mocking birds—long adapted to island life—pirate an egg.

234

235

236

237

The care of animal young is part of a long evolutionary process which, in some instances, appears to be an adaptive response to a harsh environment. Bottom-living sea animals that dwell in polar waters or in the extremely cold abysses of warmer oceans more commonly give better care to their young than do similar creatures living in other marine realms. Almost half of the various kinds of sea stars and about one-third of the sea cucumbers in Antarctic and Arctic waters either retain their eggs until they have hatched or brood those they have laid.

A Danish marine biologist, Gunnar Thorson, has found that these cold-water species produce only a few eggs or young at a time, but they compensate by breeding all year. Their marine cousins in warmer waters usually limit reproduction to brief, seasonal orgies of egg-laying, when billions of eggs are released simultaneously and without parental protection.

The female breeding in cold water is a concerned mother. She often stops eating as the time for delivery approaches. She remains still, conserving her energy as she waits for her embryos to develop. The female sea cucumber lies prone on top of her eggs or shields them with her branching tentacles. A sea star rests on the tips of her arms to hold her body above a cluster of eggs. A sea urchin holds a ring of eggs in a group around her mouth. A sea lily stands on her stalk, her arms curled over to make a brood couch.

The pressure of their harsh and icy world has forced marine creatures to push parental care to the limits of their physical powers. Some deep-water sea stars hold eggs in their stomachs and stop producing digestive chemicals until their young are ready to emerge from the egg membrane. The mother then digests the membrane so she can expel her brood into the sea. This last act is vital; without the destruction of the membrane, the young stars would be trapped inside.

Some stars and sea cucumbers place fertilized eggs in incubation sacs on their body surface, and when the young emerge they are fully independent. Others hide their eggs under a body covering near their mouth, where stiff spines with flaring tops fit together to form a false roof. Beneath its shelter is a brood chamber in which the young are completely concealed until they are ready for release. When that time comes the mother bends her body sharply, opening escape lines between the tops of the special spines, and the young move out freely.

This behavior is similar to that of the familiar lobster, crayfish, and crab. A female lobster or crayfish lies on her back while she extrudes fertilized eggs and distributes them with her walking legs across the bottom of her abdomen. They stick there to the many miniature, leglike swimmerets.

Most animals breed when the seasons are kindest to them. King penguins, which are smaller and lighter than the emperor penguins in Antarctica, breed in the summer of the Southern Hemisphere on the flat ground of subpolar islands. They are faithful parents, though they build no nest to protect their single egg. Instead, one parent stands upright with the egg, or chick, balanced on top of its feet while the other parent is off feeding. Belly skin and feathers keep the youngster secluded and safe until it is big enough to fend off the predatory skua gulls.

The emperors nest on ice in the middle of the continuous night of the polar winter, where the temperature can drop to 75 degrees below zero and tremendous gales blow. The female lays her egg, entrusts it to her mate, and then leaves for a month to feed herself. The four-foot-high father turns his back to the wind, almost standing on his heels as he balances the precious egg on his feet. He and his colleagues often cluster together to get some common benefit from their body heat, which they can maintain at about 100 degrees. They have an amazingly efficient system of converting body fat into heat. Birds at the middle of the penguin cluster conserve their energy by cooling down to around 96 degrees; in a cooperative battle for survival, they trade places with their fellows so that no emperor is too long exposed to the bitter cold at the edge of the group.

When the eggs have hatched and the females have returned, the males can leave. They stretch their legs and go off to eat, their weight far below their normal ninety pounds. The female keeps the chick warm and feeds it by regurgitation. When he has regained weight, the male returns to help feed the youngster. By this time it is spring, and the emperors have successfully bred and partially raised their youngsters before the skua gulls arrive. These gulls, rather than severe winters, are the main destroyers of unguarded chicks.

The most common species of penguin, the Adélie, and their close relatives, the Gentoos, lose more of their young to the gulls, but those that survive these hunters grow strong and healthy on the almost unlimited supply of krill crustaceans. The Adélies protect their young from the cold in quite a different way from the emperors. They leave the water in millions each October and walk about sixty miles inland over ice and bare ground to reach traditional breeding grounds, where the pebbles they need for nesting sites are available. The Adélies collect pebbles into mounds, which will later hold their eggs and chicks off the intense cold of the ground.

The Gentoos and chinstrap penguins employ another device to beat the cold. In a world with few plants, they hunt for scraps of plant material to build shelters. Neither species has much more than a foothold on the barren Antarctic continent. They are much more successful on subpolar islands where more plants are available.

The jackass penguins of South Africa throw away the stones they find as they dig steeply slanting tunnels for their nests. At their largest colony, on Dassen Island near Capetown, they have so thoroughly burrowed into the ground that in places the earth collapses under a man's foot. Instead of skuas, they must contend with predatory black-backed gulls. The jackasses have avoided Antarctic cold only to encounter the depredations of guano hunters, who have so heavily exploited their breeding areas that the birds are in difficulty. Another species, the Magellanics of Patagonia and the Falkland Islands, along with the Galapagos penguins six hundred miles west of Ecuador, were lucky in finding remote breeding areas that the guano hunters could not easily reach.

The penguins exemplify the diversity of the evolutionary process. These flightless birds are masters of the sea and can stay long months in the water. Some are strictly migratory, while others are practically sedentary. All of them use their inflexible, oarlike wings to speed along underwater, kicking their feet to steer. At sea they are fairly safe, but they must breed on land, where they face a variety of hunters from which they cannot escape in flight. The crested penguin, which lives in the fiords of the western coast of New Zealand's South Island, seeks to protect its new family in rocky caves or under the roots of trees in coastal forests. It shares the territory with the yellow-eyed penguin, a bird that goes to and from the sea by day and nests in isolated nooks among storm swept woodlands.

By separating into different species, the penguins have made an extraordinary adjustment to both equatorial and polar worlds. But they are not alone. The smaller, fish-eating penguins must compete during the southern summer with flying migrants from the north. For that brief moment, as Arctic terns flit over the penguins on their flights around islands in the Antarctic and Subantarctic, twin lines of evolution and adaptation cross paths. But the terns are only resting to prepare themselves for their great trip north to breed in the Arctic. Like the Adélies, the terns reach nest sites that may be blanketed with late snow, and they may have to scrape it away to make a place to lay an egg. Both penguin and tern show the strength of the breeding urge in their pursuit of posterity at the ends of the earth.

The common terns and piping plovers demonstrate the remarkable ability of birds to find their nests, even when they are placed in thousands of acres of beach sand where no distinctive mark is visible to the human eye. They never hesitate, dropping down accurately to feed their youngsters. A murre, wheeling over an island breeding place in the middle of the Labrador Current, comes down to a clumsy landing among thousands of her colleagues, and despite the uproar of voices, the sea of rising necks and beating wings, she lands on her own egg, which she has left lying on bare rock.

The nest seems to express the quality of its builder's life. The solitary egg lying on bare rock in the Subarctic connotes toughness and strength. By contrast, the mallee fowl of Australia and New Guinea create an almost perfect automatic nest. A large mound of soft leaves is assembled by the male bird, and as the vegetation decomposes, it warms the mallee's eggs. Some species use ordinary sand scratched over spots on the earth made hot by volcanic activity. Tending their nests, the watchful males use a variety of techniques to keep the temperature of their mounds stable; they add more material, subtract some, stir it up. Previously, they have induced females to lay their eggs in the correct place in the mound.

Cliff swallows build oven-shaped structures of wet mud plastered to overhanging rocks. Nearby are last year's nests, but experience has given the birds the knowledge that old nests can crumble.

The ruby-throated hummingbird attaches her nest firmly to a branch by building around it. The tiny cup is made with fine fibers from plants and camouflaged by pieces of lichen held in place with spider-web material and saliva. The calliope hummer nests far up the slopes of rocky mountains and must meet the challenge of the altitude; she must protect her eggs and young from night temperatures that may drop to the freezing point. In the tropics, other hummers must struggle just to get the right nesting material. Ornithologist Alexander Skutch estimates that as many as half the nest failures of Rieffer's hummingbird in the rain forests of Costa Rica are caused by the nests' collapsing when other hummers steal the supporting fibers.

The construction of the nest is vital to the survival of the species. All the factors—weather, breeding condition, availability of material and sites—must be in equilibrium if the work is to succeed. When the natural balance of the system is upset, survival is doubtful. Throughout North America and Bermuda, competition among birds that prefer to nest in holes has grown steadily more severe. Man has destroyed millions of trees that contained good knothole sites for nests; man has also introduced competitors, such as starlings and sparrows. A starling can force a flicker to abandon its nest cavity just after it has been excavated. The bluebird population of Bermuda plummeted after the European starling reached the island from the United States. Tree swallows and house wrens try to nest in openings too small for starlings, but this brings the swallows and wrens into competition with chickadees and limits their reproductive success.

The benefits of a good nest are complemented by the care bestowed on nestlings. Whether the chick follows the parent immediately after being hatched or stays in the nest for weeks, it is involved in an intensive learning process that is vital to the whole biological enterprise. When conservationists began a new program to save the whooping crane from extinction, they learned a great deal about parental care and training. Biologists studying the birds, which nest near Great Slave Lake in Canada, discovered that the survival rate of young cranes improved if one egg were taken from the nest of each breeding pair. With one offspring gone, the parents cooperated to tend the remaining chick. The cranes' behavior was different when they had two hatched eggs. Then, one adult led away the first chick while the other parent continued to care for the second. Over the years, the biologists realized that when two parents pool their efforts to tend one chick, more young birds survive than when the family is split up.

We get faint echoes of how the parental behavior of modern birds may have evolved by studying crocodiles and turtles, which have lived through millennia with little change in body form. The mallee fowl, busily rearranging his decomposing mound, is not that far distant, in a behavioral sense, from the female crocodile standing guard near the crude mound of rotting vegetation where her eggs will hatch. Turtles, which are almost as evolutionally unchanged as crocodiles, do little more than travel to a suitable place, dig their nests, drop in their eggs, cover them, and leave. Some may urinate in the hole, perhaps to give the eggs moisture in dry times.

The lack of parental concern among reptiles shows us how much more limited has been their evolution. Some species of snakes and lizards retain their eggs until active young are born, but many other species merely conceal their eggs and leave them unattended. Only a few work actively to protect the eggs. Probably the most remarkable of these concerned reptilian parents are the pythons. They are large enough at sexual maturity to produce an appreciable amount of body heat, and they coil themselves around their eggs like incubating birds to warm them. The body temperature of the brooding python is several degrees higher than that of a nonbrooder.

Venomous cobras also protect their eggs, but only the largest of them, the king cobra, which may grow to twelve feet, actually builds a nest. The eggs develop in one of the nest's two chambers while the snake stands guard in another. The cobra seemingly imitates the behavior of birds, though it has one defense no bird can equal: its venomous bite. The cobra mother will bite any intruder. In India and Southeast Asia she often lays eggs in buildings or under shrubs near paths and roads, where human beings are the most likely intruders. Most people bitten by cobras are the victims of a zealously protective mother.

The behavior of a mother defending her young is not necessarily a sign of highly evolved behavior. Venomous spiders stand guard over their egg sacs too, and the black widow and the brown recluse of the United States stay close to their untidy nests until their young hatch and depart.

The fragile egg is a helpless entity, vulnerable to cold and wet, to wind and sun, and to practically all hunters that eat eggs. Birds can afford the expensive luxury of laying eggs because they can avoid danger by flying away. The egg represents an early stage on the evolutionary road, although some cold-blooded creatures have traveled the next step, to live birth. Among the cold-bloods, parental behavior includes both indifference to the fate of the young and extreme solicitude. The billions of fish eggs floating with the plankton at the sea's surface are left to manage on their own. But some tropical fish, including the popular guppy, give birth to miniature copies of themselves. The male sea horse bumps his body against his mate until she lays her eggs in his pouch, where he fertilizes them and protects them until the young hatch and he can empty the pouch to prepare for another courtship.

The male Surinam toad clasps his mate until he has fertilized the eggs she lays. Then he spreads the fertilized eggs over her back and holds them there until they sink into special pockets of her skin. Embryonic development continues in these pockets while the female toad goes her solitary way. When the tiny toads emerge from her back, they are independent and have safely bypassed the dangerous, free-swimming tadpole stage.

The male European obstetrical toad fertilizes the female's eggs and then coils the string of jelly that encloses them around his own hind legs. He stays hidden until they hatch to tadpoles. His extraordinary transformation from ardent suitor to practically immobile egg-sitter somewhat resembles the behavior of the common water bugs of the Northern Hemisphere. The female water bug holds her mate until she is ready to lay her eggs. She cements a raft of them onto his back and leaves, while the male drops to the bottom of the pond and hides for about a month. He is not free to move until the eggs hatch and the young bugs swim away.

The young of dogfish and other sharks develop inside their parents until they are independent, and at birth there may dangle beneath them a small remnant of the once-bulging yolk sac that nourished them during their embryonic lives. Aristotle observed that the common spiny dogfish found in the Mediterranean was fed through a placenta something like that of a mammal. For centuries this observation was ridiculed by naturalists, but in the late nineteenth century Aristotle was proved to be correct. From a single mating, the four-foot-long dogfish produces a dozen pups after a pregnancy lasting twenty months. During the last part of this long, prenatal development, each pup forms a connection with the maternal tissues and gets additional nourishment after its yolk supply is exhausted. At birth it is an active, independent creature about twelve inches long.

The diversity of parental care forms a mosaic in the slow order of evolution and gives us vignettes of life operating at different levels. When, during the nineteenth century, the marsupials and monotremes entered the literature of natural history, many naturalists and evolutionists were confounded by their odd habits. The female spiny anteater, a monotreme, could incubate her egg in a temporary pouch on her belly surface. The marsupial mother could travel about freely with her young attached firmly to her nipples in a permanent pouch on her body. The traditional naturalist might have been baffled by these odd creatures, but if he studied them closely, he might have seen how well they confirmed Darwin's theories of universal evolutionary laws.

It was true that the duck-billed platypus laid eggs, but then, like a bird, she kept them warm, by curling around them because she had no pouch. And the shrewlike coenolestids of South America were marsupials in every detail, except that each female lost her pouch as she matured. The cat-sized American opossum was equipped with a well developed pouch and gave birth to about twenty youngsters every year, each one about the size of a honeybee. Yet only thirteen of them survived, one for each nipple in the mother's pouch; the rest were lost as they searched desperately for a place to nurse.

The birth of young opossums occurs less than fourteen days after mating. The mother leans back against a support and wets the fur between her vaginal opening and her pouch with her tongue to smooth the road for her youngsters as they haul themselves blindly toward the pouch. Their front legs are well developed and have strong claws, but their mother may use her pointed nose to help open the entrance to her pouch as each youngster falls in.

Inside the pouch, the young must find the mother's slender nipples, and each youngster must work the tip of it down its throat into a tiny stomach. There, the tip of the nipple expands slightly to form a button, which prevents the youngster from losing its grip. At suck, the young opossum increases its weight tenfold in the first week, to about one twenty-seventh of an ounce.

After about twenty-eight days, the youngsters' throats have enlarged enough to release the nipples, and the young are ready to look out over the rim of the pouch and see their world. In another seven days they are ready to climb onto their mother's back. She helps them stay in place by curling her tail up over her back. Each youngster uses its own prehensile tail to cling to hers while it grasps her back fur with all four clawed feet. Eight weeks after birth, the young are weaned and leave their mother. She may mate again and raise a second family before the summer ends.

The newborn kangaroo weighs about one ounce and must make the long journey to its pouch with no help from its mother. Unlike the opossum, the doe kangaroo does not lick a helping path for it, though she does clean up the moisture left behind by her clambering baby after the birth, and she may temporarily prevent her joey, the older, ambulatory youngster, from suckling.

The doe kangaroo almost always has two nursing youngsters of different ages. She is able to produce a small quantity of extremely rich milk in the slender nipple to which her newer infant is attached, while a larger quantity of a more diluted milk is available, in one of the four nipples in her pouch, for the joey, which follows her but is too large to fit into the pouch. Though the joey is slow to wean itself, it is usually independent before the pouch kangaroo frees itself from the slender nipple and appears over the edge of the pouch.

On the great Australian plain, where droughts sometimes last for years and torrential rains transform the landscape overnight into a jungle of wild flowers and grasses, the red kangaroo uses variations in pouch-birth routine to survive. When hard-pressed, the mother summarily ejects and abandons her pouch youngster. But as conditions improve, she can call upon a dormant blastocyst—what might be called an embryo-in-waiting—which is there for just such an emergency. This new youngster reaches development rapidly and is born within a few days. The mother is then free to mate once more and create another embryo, which goes into dormancy to await the call for birth. Even though the doe kangaroo finds it difficult to secure enough food to sustain herself, even though the pouch youngster and the joey die, the dormant blastocyst remains alive, drawing very little from the female's life force.

Such embryonic pauses are nothing exceptional. As a technique, it has proved advantageous to some bears, weasels, seals, armadillos, bats, and roe deer. It allows parents to mate at convenient times and the young to be born when the mother—and her offspring, in their turn—can most easily find food.

As any young creature approaches independence, it faces a complex world of new sights, new sounds, and a diversity of foods. Somehow, it must respond to all this, either by watching its mother, learning by trial and error, or by acting upon its inherited tastes, which tell it what to eat and what to avoid. The young pouch kangaroo leans well forward between the legs of its mother as she nibbles at soft new buds. The youngster gets a grandstand view of food possibilities and, when it is older, it will copy its mother's eating habits. The koala bear, an Australian marsupial, is born like any other pouched mammal. But the mother's pouch opens to the rear instead of the front. When the young koala is old enough to peer out of the pouch, it looks between the back legs of its mother and cannot see what its mother is eating. But the mother's diet of eucalyptus leaves is not completely processed, and this partially digested excrement is the young koala's first solid food. Soon, the baby matures enough to ride on its mother's back, where it is able to reach for fresh leaves when its mother is feeding. The young koala already has an indelible taste for the correct kind of eucalyptus leaves— fewer than half a dozen species among six hundred.

A young starling will reach into a large bowl containing a hundred different species of insects and deftly pick out sixty species, leaving forty uneaten. Some of the ignored forty so closely resemble those that have been eaten that a human observer cannot tell them apart. But the young starling knows.

Young African wild dogs wait in the care of females at the communal den for hunting parties to return with food. The adult hunting dogs have eaten, gulping down great chunks of wildebeest or zebra, gazelle or impala; the meat has been partially digested during their return trip to the den. As they meet the hungry youngsters, the adults begin disgorging the meat while the young dogs yap and clamor at their mouths. If the young are too small to accept the partially chewed meat, the older dogs will eat it again, chewing it more thoroughly and digesting it further before disgorging it once more. The ritual is continued until all the young dogs are fed. Only then are the hunting dogs free to digest the rest of the meat for themselves.

In the rain forests of Panama, a female two-toed sloth crawls laboriously across a clearing, her body dragging on the ground, her small brown youngster clinging to one side. The mother scrapes a bush, and her youngster falls off. For almost a minute, the female sloth continues on her way until she is about one body length from her youngster. The baby sloth opens its mouth but no cry sounds to the human ear. The mother stops, opening her mouth in imitation of the youngster. They are probably communicating in ultrasonic frequencies. The dialogue continues for nearly five minutes while the youngster crawls around the shrub. The moment it catches its front claws into the fur of its mother's back legs, she resumes her slow march toward the nearest tree. It would appear that the sloths have solved one nursery problem, but communication for them is possible only within a limited range. If the young sloth had been out of earshot, the mother might not have been able to return along the line of her own scent to search for her youngster.

A lioness, on the other hand, can growl her cubs into hiding while she bounds for yards along the banks of a fast-flowing river to rescue a youngster that has been swept away. When she returns with her wet cub, the others are where she left them, and the family is reunited.

Elephants, hippopotamuses, whales, and porpoises also show extreme solicitude for their young. A hippo mother tries to keep her newborn baby at the water's surface where it can breathe until it learns to get air on its own. She holds the youngster right side up and will continue to do so for some time, even if the youngster dies.

The realization that parental behavior is genetically programmed may be uncomfortable for the layman to accept, until he realizes that what seems to be a mechanical response allows animals to act in extraordinary ways. The behavior is always within predictable limits. The creature does what must be done to serve the species best in the struggle for survival. Solicitude for the young performs a vital function that human beings can understand, but its complexity in certain animals tends to be baffling. The extraordinary interaction between African hornbill parents and their young, for instance, has only recently come to light. To begin with, the birds have trouble finding a suitable tree cavity in which to start their family. Once they have found it, and mated, the female hornbill enters the hole and helps the male to seal the entrance with mud. Only a very small opening is left; through this he pushes food while the female incubates her eggs and broods her nestlings. When they are old enough for their mother to leave them, the parent hornbills peck open the door, and the female emerges. The birds seal the door once more with mud, but this time they are helped by the two or three young birds inside. Both parents now bring food and push it through a gap too small to admit any creature that might harm the nestlings. Only when the youngsters are ready to fly do they chip away the hardened mud. This cooperative system depends on the survival of the male bird, but even when both parents are feeding the larger nestlings, the death of one so upsets the delicate balance that some of the youngsters may die.

Safety for the young is secured by far cruder methods in any large nesting colony of sea birds, though cooperation is part of the program. The moment a predator appears, the colony explodes into a blossoming umbrella of excited adults, and a passing eagle will find itself the target of diving birds, showers of half-digested fish, and excrement. This cooperation preserves all the birds.

In tropical America, the anis, which are members of the cuckoo family, have taken the concept of social nesting to the commune level. A female ready to lay an egg deposits it in any unfilled nest, and any adult or juvenile seems capable of incubating the eggs and bringing food to nestlings. The entire flock bands together to repel intruders. Ostriches also cooperate in rearing their young, and the male takes turns warming eggs that several hens of his harem may have laid in a common nest.

Cooperation to preserve the young is by no means limited to birds. When danger threatens, bison and musk oxen of both sexes, their horned heads facing out, close ranks to form a circle around their young. Elephants and most wild primates go to extreme lengths to protect youngsters in the group. Frequently, older animals of either sex remain with the mother during birth, and often they expose themselves to danger to draw attention away from a youngster in difficulty. Certain lactating bat mothers allow any young bat of their own kind to nurse.

The concept of cooperation can be seen at work among the simpler forms of life. Sexton beetles and scarabs work together in pairs to bury supplies of meat or balls of dung to sustain their larvae. Adults and juvenile male termites work with their female counterparts in the dark galleries of termite nests for the good of the colony. There are no termite drones, although the social group includes some creatures with large jaws and heads which act as "soldiers," and others that specialize in spraying formic acid to defend the colony.

Some of these so-called simple lives are capable of what we see as sophisticated parental concern. Until quite recently, it was assumed that when the fertile females of some solitary bees and wasps built a store of food to sustain their larvae, their concern ended with the laying of the egg and the closing of the chamber that contained it. In the minds of men, the mud dauber wasp filled her earthen cells with spiders, and that was the end of it. The sphex wasp stored paralyzed caterpillars for her young and forgot them. The bembex wasp stocked a burrow chamber with flies and went off. The great sphecius wasp carried away large cicadas for her larval young and then moved on. But now it appears that some of these creatures return occasionally to check on their developing brood. The alkali bees of the western United States come back and reopen any cell infected with fungi. They pack the infected space with soil, and so block the normal production of fungus spores.

The common bembex wasp plugs the burrow opening each time she adds another paralyzed fly to the collection in the end chamber where her egg is to be laid. When she returns with another fly, she digs out the plug and stores the new food. When she is finally stimulated to lay her egg, she no longer fills only the burrow mouth. Instead, she packs the entire tunnel. To get materials, she digs accessory burrows near the entrance. Each of these resembles the original burrow but in fact is an empty cul-de-sac. She eventually obliterates the true entrance, leaving the accessory burrows open. By this diversionary tactic, the bembex distracts parasites searching for her store of food. They may repeatedly explore the accessory burrows but soon depart after learning that each is empty.

The outwardly simple bumblebee nest contains a highly developed society in which the young are carefully protected. Each underground queen builds a simple nest floored with fine plant fibers. On this surface she piles fresh pollen stuck together with nectar or honey. Nearby, she constructs a waxen honey pot and stores liquid honey in it. Then she is ready to lay eggs on the pollen pile. When this is done, she covers the eggs with a thin sheet of wax to form a brood cell.

The eggs hatch into maggotlike larvae, which feed on the cell pollen for about fourteen days. During this time, the queen pierces the wax cover and through the hole regurgitates honey mixed with pollen as the youngsters lick her mouth parts. After each feeding she reseals the waxen cell. When the larvae stop feeding, they spin their own thin, tough cocoons of silk and start transforming into adults. About fourteen days later they emerge as unmated females, or workers, somewhat smaller than the queen.

The real primitives among social insects give us some insight into the range of behavior in parental care. Australia's bulldog ants are true primitives. These ferocious, amber-colored creatures are almost an inch long. They search through the leaf litter, seize insects and snails, and invade termite nests to capture prey. Their powerful mandibles and great strength enable them to carry fresh meat back to their nests where comrades are guarding the developing young. There, groups of white eggs lie on the bare ground. Squirming larvae and separate heaps of cocoons are also often exposed. The hunting adults step over the active larvae, drop unregurgitated food fragments, and step back. The voracious larvae gorge on the meat and fight each other for it. The bulldog ant's parental care has a frontier quality about it; the ferocity of their defense of the nursery area is unique in the insect world. Such fierce defense apparently makes up for the lack of underground fortresses that other ants and termites build.

The driver ants of Africa and the army ants of tropical America are also primitives, but they are more programmed than the bulldogs, and therefore more predictable. Most of their time is spent wandering in nomad columns in search of booty, but at certain seasons this movement is controlled by the need for parental care. The queen halts the column when her abdomen becomes so distended with ripening eggs that she cannot keep up with the rest of the army. During their travels, the workers have been carrying and feeding developing larvae. As these larvae achieve full size, they stop regurgitating a substance that stimulates the workers to move. Now, with the whole column halted, the larvae spin their cocoons and begin to pupate. Meanwhile, the eggs of the queen mature and she lays as many as 25,000 in fewer than seven days. The eggs hatch, and the young larvae begin to secrete the substance the tireless workers need to be stimulated to travel. The queen is active enough to march again, and the column once more moves off in its nomadic search for food.

Parental care is a cornerstone in the great edifice of behavior. The interaction between adult and youngster gives a high value to the role of the mother and lays a base for the transfer of learning. It rewards those species that protect their offspring by giving longer life to greater numbers. Many of the signals which release the panoply of parental behavior are still hidden from human eyes, though more is being learned every day. The interplay between adult and juvenile is a labyrinth of stimuli—chemical, audible, visible, tactile—passing from one animal to another. The signals keep the enterprises coordinated, the schedules of life-fulfilling production continuous.

133

Safety in Numbers

Photographs by:
Fred Bruemmer, 252, 253
David Cavagnaro, 251
Arthur Christiansen, 255-259
Stephen Dalton/Natural History
Photographic Agency, 241-246
Francisco Erize, 254
Keith Gillett, 248
Mervin W. Larson/Bruce Coleman, Inc.,
238, 239
Rolf O. Peterson, 260-264
G.A. Robilliard/Sea Library, 250
Ron and Valerie Taylor, 247, 249
Karl Weidmann, 240

Large flocks of starlings or grackles or swallows, darkening the air in their massed flights, display animal behavior that has many meanings. Obviously, there is safety in numbers. Massed colonies of seabirds raise such a fuss when an eagle passes nearby that the big bird of prey is often intimidated and flies off. A group of African antelope or zebra are more difficult to kill en masse than when they are alone. Powerful muskoxen, masters of the Arctic country, gather in circles to protect themselves against wolves. Animals in groups, particularly fish, trigger alarm systems, instantly communicating danger to every member of the group. Dense-flying birds all change direction simultaneously at the transmission of some unidentified signal. Animals crowd together for warmth in winter. The emperor penguins of Antarctica use the combined mass of their bodies to break up the force of the wind so that each individual is a little less chilled.

Animals migrate in groups, and there is less chance of young individuals losing their way. Young seals, seabirds and fish gather in groups of their own age in massed effort to intimidate or confuse predators.

But the grouping of animals has many other advantages. It enables dominant animals to exert influence over the less dominant, either by example or by the use of fighting or threat. The pride of lions functions more efficiently than nomad lions, not because of the dominance of the male lions necessarily, or because of their size and strength, but because the lionesses cooperate to hunt. The lions fight to preserve the territory of the pride, and this is for the benefit of the group.

Some animals, and particularly ants, that live in colonies have evolved specialized behavior and even specialized bodies as a means of helping all the members of the group. In the southwestern United States and in Mexico, certain "honey" ants collect the sweet liquid exuded by galls on oak trees (238-239) and take it to underground storage chambers where colony members serve as storage vessels. With their bodies swollen and made translucent by the amber liquid, these ants quietly yield "honey" upon demand or accept more for community storage. "Leaf-cutter" ants of the tropics of Latin America (240) cut pieces from fresh leaves and, marching in line, haul the fragments home to the compost heaps on which grows a fungus that feeds the ant colony. Both termites and ants have developed specialist castes of their kind to do different work together.

238

239

240

The operation and the continuity of a honeybee colony centers around a single, long-lived queen, the only mated, fertile female in the hive. Some signals induce her to lay unfertilized eggs that hatch into drones, or males. A drone, with enormous eyes and a body somewhat larger than the workers around him, solicits food from the workers on the surface of the comb (241). The queen lays her eggs in a brood comb, one to a cell (242). They hatch into maggot-like larvae after a gestation period of only a few days, each one curled up in its cell and fed with food regurgitated by the workers (243). In this pupal stage (244) the honeybee transforms from a larva into a small version of an adult bee. When mature, it must break through the sealed entrance of its cell in the brood comb (245). At the opening of a hive in a hollow tree, a field bee is airborne (246) and ready to go to work.

241

242

243

244

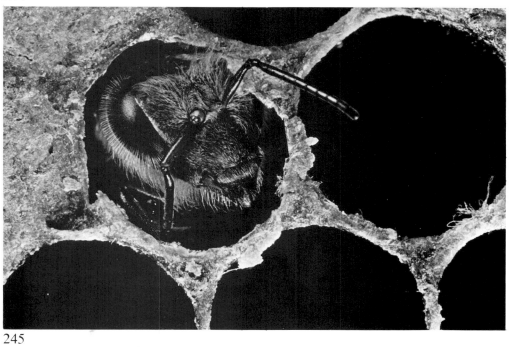

245

246

249

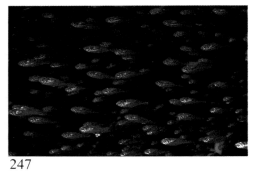

247

250

248

251

Many kinds of fish travel in schools. These are swimming in the coastal waters off Australia (247). Goose barnacles settle together for life whenever they can find a suitable site, such as driftwood (248). Anemones also spend their lives fixed in cases stuck to rocks of tidal pools. Each return of the tide brings them small food particles which they capture by extending soft tentacles studded with nettling cells (249). Sea stars of the genus *odontaster* cluster together at depths of 30 meters or more (250). Pelagic crabs swim together for common protection (251). Sleeping walruses pack together for warmth and comfort on the hard rocks (252). Alaskan fur seals congregate on rocky shores where dominant males defend their harems (253). All of the royal penguins in the world congregate for their breeding season on Macquarie Island, halfway between Australia and Antartica (254).

252

254

253

Territorial Display

The male of the large sandpipers are known as ruffs, the female as reeves. The sexes look alike during winters that they spend in northern Africa and southern Asia. When they fly to northern Europe and Asia in the spring, each male grows a spectacular collar of stiff feathers resembling the ruff of Renaissance clothing, and courting begins. The mating grounds are hillsides and meadows the birds have used for generations. There, they display, the ruffs dancing, posturing, pushing their beaks into the ground and quivering. The reeves enter the mating grounds individually to select a mate. Afterward, the female builds a nest, lays eggs, incubates them, and tends the young (255-257). The spectacle of the ruff in full courting dress is the genesis of the entire act (258-259). Many mammals and birds use elaborate rituals involving play and mock fights before mating.

255

256

257

258

259

The female wolf, even when she is in heat and eager to mate, defends herself from any pack member that makes a threatening gesture (260). The male who is eager to mate with her, but who may appear threatening by his larger size, seeks to diminish himself by rolling on the ground, as though he were only a puppy. She responds with play gestures that resemble her behavior when caring for her young and also when she is courting (261). When an aggressive but subordinate young wolf attempts to join the courtship, the older male attacks and thrashes the youngster (262). He, in turn, rolls over on his back and thus avoids a fight by his submissive behavior. A subordinate male (with white fur) who approached the leader's mate is punished by the leader (264). The pack often plays together in simulated attack and defense (263).

260

261

262

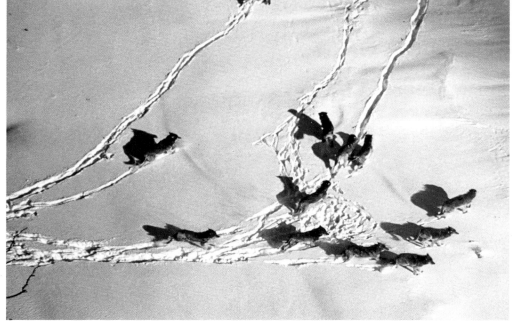

263

264

In the depths of Mammoth Cave in central Kentucky, blind cavefish wait out their solitary lives along the underground Echo River, which flows through the cave. Their three-and-one-half-inch-long bodies are pinkish-gray and show only a faint trace of blood. They look like minnows, and, like minnows, they are sensitive to the scent of others of their species, but they do not form schools as minnows do, since they lack the necessary sense organs. Alone, they are at the mercy of their environment. If the river temporarily dries up, they starve. If the river floods, they are swept to the surface where, blind and helpless, they become easy marks for predators.

In contrast to the solitary lives of the cavefish are those fish that school. Some come together for short periods; others stay in schools much longer. Each fish must be immediately sensitive to the movements of the others around it and must be able to coordinate its own actions to the rhythm of the school. Every fish maintains a set distance from nearby fish. This living space shrinks only if danger threatens; the close-formation swimming of the school will confuse a hunter, making it uncertain about which individual to attack. But if an enemy does attack, the fish scatter abruptly, their behavior changed by the flavor of an alarm substance that has been discharged. The attacker is given no chance to follow the school and pick off its members one by one. When the enemy has disappeared, the fish reassemble to carry on their group swimming. The group "family" is vastly more conspicuous than a solitary fish, but the lone cavefish is much more vulnerable to any enemy that can reach it. The group creature has a better chance of finding food with its fellows, but it must compete for the available food. The benefits and disadvantages to creatures that live together seem so evenly balanced that other factors must tilt the scale, which they do. The group ethic transforms the behavior of individuals in ways that are still not completely understood. We do know that many kinds of birds and mammals eat more often and in greater quantities when they are stimulated by the group. Frequently, group members mature earlier, and throughout their lives in the extended family they all benefit from one supreme advantage over the loner—the opportunity to learn by imitation.

Of all the creatures capable of gathering together for mutual protection, for migration, or for breeding, birds are the preeminent examples of togetherness. They vividly demonstrate the nature of the extended family, showing its power and strength when they come together. Mute swans leave their New England millpond in December to winter along the shore, where salinity and tidal flow keep waters open during the winter. In March the birds are back on their millpond to rebuild the old nest or begin a new one. They still accept their young of the previous year, but the acceptance wanes as the youngsters approach ten months of age. The family benefits from winter flocking, but the mute swan parents will not tolerate interference with the new season's breeding, so they cast out their youngsters, which scatter to find companions of their own age and settle where no other mute swans are trying to raise a brood of cygnets.

When birds nest in large colonies, the young commonly gather in groups while their parents are out hunting for food. In these nurseries they appear to lose identity until their parents return, whereupon individual cries reunite families and food is disgorged. The gathered young have no real defense against their enemies, but their numbers intimidate prowling eagles, falcons, or gulls. Skuas are expert at seizing any young sea bird that strays only a few feet from the mass of its fellows.

The restricting of the sex act to only those males which are the finest specimens of their kind is one of the basic biological reasons for the extended family. Among many male birds, dominance must be reestablished every year, and practically the sole function of the males is to earn the right to mate. A snipelike bird, the ruff (the female is called the reeve), which lives in the tundras of northern Eurasia and in damp meadows somewhat farther south, demonstrates this need in extraordinary fashion. The male-female bond is almost nonexistent, since the sexes are together for only a few minutes each year. The rest of the time, they flock as separate sexes.

The ruffs return from wintering in Africa and Asia ahead of the reeves. They move to their ancestral courtship areas, where the dominant ruffs repel other males as they claim their tiny territories. When the reeves arrive, they walk among the demonstrating or dozing males, inspecting prospective mates. A reeve lets a ruff know he has been chosen by extending her beak and nipping him. He leaps forward and mounts her at once. In a few seconds they separate and the reeve walks off or flies away. The male is equally uninvolved and may, indeed, go back to sleep.

The existence of a peck order to establish the dominance of one individual over another functions at all levels of animal life, though it can be seen most clearly among domestic fowl, which have separate peck orders for the cocks and the hens. Cock Number One may peck Cock Number Two without starting a fight, and he is more or less obliged to peck his subordinate several times a day to reinforce his dominance. The second cock in the hierarchy can peck any of the others, except Number One. On the female side, the dominant hen chooses the best place to nest and dominates all the other birds of either sex, except for Cock Number One, to whom she defers at the drinking trough and food tray. The peck order is maintained by many subtle checks and balances. The prime male, for instance, can mate with any hen, but he may shun the dominant hen if she is too aggressive. Frequently, far down the peck order, there may be a bird so thoroughly hen-pecked that the dominant creature can ignore it completely. This low-caste individual is always excessively submissive and may insinuate itself into the dominant's presence, eating and drinking with it while the others stand respectfully at bay.

The idea of a peck order among birds originated with a Norwegian psychologist named T. Schjelderup-Ebbe, who worked with domestic fowl and more than fifty other kinds of birds, but it was the distinguished behaviorist W. C. Allee, of the University of Chicago, who advanced the theory that dominance was essential to the survival of social animals of all kinds. Convinced of this fact, he gathered as much evidence as he could to show that dominant-submissive relationships give a necessary balance between disruptive and cooperative forces in animal societies.

The control of aggression is so vital, in fact, that dominant creatures reinforce their status by stopping conflict among others.

More recently, John Price, of the Institute of Psychiatry in London, has suggested that modern man still lives with the remnants of a peck-order system that comes down to us from humans who congregated in small groups thousands of years ago. There had to be a strict hierarchical order then, he thinks, to assure stability and survival among these early men. Each individual knew all the others in the group and accepted his or her place in the peck order, thereby cutting down unproductive competition and unnecessary fighting. But today human beings move from place to place and from group to group. Gifted individuals enjoy the elation that accompanies exceptional success, while many others resent their lowly social status in communities where too many strangers must live too close together. Price wonders if this situation might not be one of the basic reasons for the soaring incidence of mental illness. It is a question that cannot be disregarded when we look at how other social mammals have worked out their group ethics to ensure survival and stability. The groupings take many forms; some do not even faintly resemble a family-type arrangement.

Animals in bachelor herds are special adjuncts of the extended family. These males, some of them not yet fully grown, gather near nursing females and their young. Bachelors—whether they are antelope, horses, or fur seals—are kept away from the available females by the dominant males.

Dominant fur-seal bulls come ashore early in the breeding season to claim territory and to settle boundary lines. They recognize one another and meet frequently along the borders of their territories. There is practically no need to fight, except to evict aggressive bachelors. The territorial bulls work to keep the bachelors away and inspect every female that comes ashore in their territory.

The bull knows individual members of his harem as they return from fishing expeditions to nurse their pups, and he ignores them. He is waiting for a pregnant female, knowing that she will give birth within a few hours of her arrival and will be ready almost immediately to mate. Once he finds his female, he stays as close to her as possible until she is ready to mate. During all this time he does not eat, but neither do the bachelors, which have been restricted to higher ground. Unable to eat, determined to defend his territory, and preoccupied with the need to mate, the harem master spends up to sixty days in exhausting shore duty. By the end of his shore time, he is weaker than the waiting bachelors and returns to the sea to regain his strength. The members of his harem and their pups leave too, though he now has no interest in them, nor in the newly arrived virgin females. As soon as he leaves, the latter take over the land and the bachelors swarm down from the high ground to meet and impregnate them.

The mountain sheep of North America represent another form of the extended family. Creatures of a continuous frontier, the rams live high in the mountains, feeding off rocky slopes well above the ewes until winter storms force them to descend. On their way down, the rams fight on small arenas of level ground to establish the order of their dominance. The rams are like duelists; they turn away from each other and prepare for the charge. Just before the tremendous collision, the rams lift their heads and then the massive horns slam into each other. The animals drop to all fours but no great damage has been done despite what appears to be a stunning impact. The dominance of one ram over another has been established, and no animosity lingers after the collision. From these encounters comes an order for all the rams. A few must be dominant, and thereafter the dominant animals need only threaten those beneath them to reinforce the position they have won.

But Valerius Geist found in intensive studies of these sheep that the powerful rams were considerably less dangerous than the ewes. The rams often get down on their knees and bleat like lambs when they approach ewes. This infantile attitude reduces the ram's enormous size and helps to prevent the aggressive female from charging him and perhaps wounding him with her sharp horns. Geist found out that immature behavior is common among mountain sheep and members of the deer family living at high altitudes. In fact, the males of these two groups act like human teen-agers in their rowdy passage to maturity. Geist believes their boisterous behavior is the result of their living at high altitudes through the ice ages of the past two million years. They had many opportunities then to extend their ranges into virgin territory in the wake of retreating glaciers, and could feed on plants that grew quickly and compactly, concentrating into their foliage the abundant mineral nutrients left behind by the glaciers. On such a rich diet, the plant-eaters became large and matured sexually at a younger age than similar creatures at lower altitudes. It was natural, Geist feels, for such animals to behave like adolescents because they inhabited what amounted to a constant frontier. Their great strength and adaptability was better suited to survival than rigid conformity to set patterns of behavior. It is no surprise to Geist that the biggest and most powerful of all the moose still roam the cold bog forests of Alaska. Huge animals of other species prowled the fringes of the ice-age glaciers until men crossed the Bering land bridge from the Old World some twenty or thirty centuries ago and moved into the same territory.

The fierce preservation of a hierarchical order is not confined to these larger mammals, nor are territorial claims. European rabbits that were brought to Australia years ago mark their home territories and defend them from invasion by rabbits from other colonies. Each rabbit warren has its dominant buck and dominant doe, and their offspring are most likely to inherit the best central burrows of the warren and raise to maturity more young than the other inhabitants. A prairie dog colony also has its strict territorial limits, and each generation learns to defend this territory so it can be passed on to the next. Tassel-eared squirrels living high in ponderosa pines in the American West fight to establish hierarchies and defend their territories. Mice not only battle for dominance but also help to preserve their positions by biting off the facial hair and clipping short the whiskers of low-status creatures to mark their place in the hierarchy for life.

Like many primates, baboons pool their abilities in order to survive as a group and to obey the commands of their leaders. They are fascinating to watch as they work their way through eastern Africa south of the great deserts in troops that range from five to about two hundred. For them, the welfare of the group is more important than the independence of the individual, and they express in their behavior the essence of the extended family.

The world they live in fosters togetherness; they seem so vulnerable fossicking along through the grasslands, picking away at seeds, dead insects, traveling ants, and termites. At night, to protect themselves from their great enemy, the leopard, they sleep in extremely high trees or along precipitous cliffs. There is, therefore, a positive need for them to control their squabbling if the safe place is crowded, and to rest quietly together against the common danger. Once the sun is risen, the animals usually split up to travel in smaller groups, but they remain tied to the security of the sleeping place.

In the arid territory along the border between Kenya and Ethiopia live baboons once venerated by the ancient Egyptians. Hunting this desolate land in small family groups, each led by a large male, they move from one stunted acacia tree to another in search of scarce food. Farther south, where there are taller and more numerous trees and some cliff refuges, the country is kinder to baboons, and they gather in larger groups. An average troop has about forty members, and while several troops may mingle at water holes, each maintains its independence. It is probably impossible for a baboon to switch troops, unless some large-scale disaster forces changes in behavioral patterns.

To the human watcher, a baboon troop seems disorganized and chaotic. Adults wander about picking up small plants, seeds, and insects while infants cling to the belly fur of their mothers and older babies ride on their backs. The young snatch food from their mothers, preparing themselves for the moment of weaning, and juveniles play quietly together in small groups. But if the stillness is broken by a cry of pain or a squabble, at least one male baboon immediately appears to restore peace. Otherwise, the big males and mothers with youngsters sit quietly among their subordinates, which work over their dominants' fur with fingers and teeth. The highest-ranking baboons, usually the largest and most powerful, sit at the edge of the group, still as statues. As if to show their disdain for the lesser animals, they frequently face away from the crowd.

Meanwhile, young males of lower status prowl the outskirts of the troop, and if one discovers food, sees an unfamiliar object, or senses danger, he yelps for attention. His calls attract the big males, and together they set out to investigate. Few hunters dare to face such a massing of baboon strength, and the troop is well protected by these guardians. During any confrontation, the other members of the troop are likely to move off in the opposite direction. When the danger is past, the big males shield the retreat. They rarely lead the troop; their job is to defend it. The juvenile males move in advance of the troop and are the most vulnerable if a leopard chooses to ambush one of them before the victim can summon help.

During long periods of tranquillity the baboons test the social structure of the troop and make adjustments if they are necessary. The dominant male need only yawn, showing his large and menacing canine teeth, to threaten every other baboon nearby. If these neighboring baboons do not move away when he yawns, he many charge, and he will certainly attack if a subordinate animal has the temerity to make a threatening gesture in his direction. He is tolerant of subdominant males as long as they display the correct submission and respect for his authority. These subdominants, in turn, threaten individuals lower than themselves on the hierarchical ladder, though they are careful to face away from the dominant male when they do so. In this way, the chain of threats passes down through the troop until the intimidation reaches the most subordinate baboon and peters out. All this testing ensures that the most vigorous males remain close to the chief and that there will always be a second-in-command ready to step into the leader's position.

Within this power structure, relationships develop that indicate at least part of the strength of the extended family. The lordly males appear to be totally uninterested in their inferiors, but this is untrue. They may have close ties to these animals, and they are particularly concerned about the welfare of mothers with youngsters. If a female with a newborn baby lags behind the traveling troop, one or several males may slow to flank her and thus offer her and the baby protection. And yet an inferior male that falls behind because he has been injured or is sick is absolutely ignored; not even his screams of anguish affect them enough to come to his rescue.

Much credit for such detailed and fascinating information must go to Irven De Vore, of the University of California, who has observed baboons close-up in equatorial Africa for many years. He has recorded their exceptional visual alertness and has found that their attention-getting sounds come close to language, though they use other signals as well. A female wanting to mate approaches a dominant male and displays her swollen sexual area to get his attention. This is almost exactly what a subordinate male does to appease his superior and to acknowledge his own inferiority. The dominant male often mounts him momentarily, as though he were about to mate with him in a homosexual manner. But it is only a signal, meaning that one animal accepts the superiority of the other without fighting over it. Indeed, if an inferior female presents herself to a dominant female she, too, may be mounted for a moment. This use of body language sometimes lets inferior animals get close to mothers with infants, and if the signals are right, the mother may allow the submissive creature to touch her infant.

Within the hierarchical baboon structure is another, longer-lasting pyramid of domination. The chief males maintain close relationships with the chief females; both inferior males and females are kept at a distance. The same separation is maintained down through the troop so that social differences tend to be perpetuated. The youngster of a superior female most likely will grow up in the power center of the troop. If he is male, he has an excellent chance to mature as a subdominant animal with a strong possibility of becoming a top-ranker. A female has a good chance of perpetuating the position of her mother. There is a strongly entrenched power elite, and the only way it can be penetrated is by an animal who uses subterfuge to gain entry. This baboon is especially submissive and anxious to groom, to baby-sit, or to be a "servant" to superior males and females. Gradually, if the baboon has sufficient guile, it will gain admission to the club, its membership earned by merit rather than by family connections.

Little about the extended family is "democratic" or "fair," since the hierarchical principle is only concerned with what works best, and all primates use it in one way or another.

Early in the 1950's, a group of macaque monkeys was kept under observation by Japanese zoologists on the island of Takasakiyama. The monkeys slept on the upper slopes of the island's mountain and came down in the morning to feed from supplies provided by the scientists. The young males always walked in front and at the sides while the dominant males and females with infants traveled in the center of the group. Rank dictated which monkey ate first. When they rested afterward, the dominant macaques, along with females and young, chose the most attractive area to relax. No subservient males were allowed to come near, and any animal that dared was severely punished. So strict was the hierarchical order that lesser animals fled at the approach of a dominant male.

To test dominance, the scientists would toss a bit of food between two males to see which one would grab it. Even if the food rolled between the legs of the lesser animal, he would not touch it. If the dominant animal had eaten his fill, he could not let the lesser macaque eat the food without first mounting him to assert his authority.

The tail of the long-tailed macaque monkey is an integral part of the caste system. Mature monkeys let their tails droop as a signal to other members of the family that there is no need to defer to them, but the dominant male carries his tail arched above his back. He disrupts matings if he does not approve of them, and may mount the female himself. His position is so secure, in fact, that he may father most of the young in his territory.

The power of dominant gorillas is much more subtly exercised. Among the lush forests of bamboo on the high slopes of dormant volcanoes along the eastern rim of the Congo basin, animal behaviorist George Schaller found ten family groups of gorillas living in close proximity. As a result of long and patient field work he was able to ingratiate himself with the two hundred gorillas making up the ten families, but he quickly learned to look away when a large male stared at him, since the prolonged look could be as much of a challenge as a physical attack.

Schaller called these full-grown males "silver-backs" because of the graying hair on their backs. He discovered they sometimes weighed as much as the three-hundred-pound adult females. A dominant male wielded such power that he could influence his family merely by standing up, facing in the direction he had chosen to move, and then looking at each member in turn. If any one of his subordinates failed to respond to his message, the big male needed only to tap the animal with his finger to get its attention before repeating the silent order. Then the family walked off slowly together to find new groves of young bamboo shoots.

Schaller followed the gorillas for many hundreds of hours and found that they defended no territory, moving instead through a roughly delineated home area. Because they were almost constantly on the move, they rarely bedded for the night twice in the same place. When one family met another in a contested area, the silver-backs, keeping their distance, glared at the rival animals while females and juveniles walked restlessly about waiting for the order to move. Once the silver-backs turned away, they obediently followed. Only the young males, who weighed about four hundred pounds, seemed to be free of the family ethic and could wander through the forest to join another group if they wished. To gain a place in the new family, they stayed on its outskirts for a day or two, and were always submissive to the staring challenge of the silver-back in charge. Once he was accepted, the newcomer was often allowed to mate with a receptive female.

The actions of the gorillas were quiet and undemonstrative until a silver-back became excited, perhaps by the presence of a man, and hoot-barked a powerful call that silenced his family. Rearing up, he broke a branch and gave out a full roar as a prelude to beating his cupped hands against his chest. The sound was like the tattoo of a low-pitched kettle-drum. The enraged silver-back jumped about, picked up sticks and threw them at the intruder, ripped off foliage, thumped the ground as well as his chest, and ran back and forth. Finally, he stopped and returned to his first challenge: the long, steady stare.

Aside from these alarming outbursts, the gorillas that Schaller observed were generally quiet and not aggressive. They liked contact with each other but they had never developed physical caresses into a ritual or a routine. They did very little mutual grooming, though most other primates are very fond of it.

The notion of the elite animal's perpetuating himself through countless generations is so disturbingly close to much of human behavior that many researchers have had to ask themselves whether or not elitism is in the normal nature of things. The extended family of the wolf pack, for instance, in which the dominant animal takes what he wants for himself, depriving the others, resembles those human cultures in which the rich man has more wives and children than the poor man. Yet we humans cannot claim to inherit an ironclad system of elitism developed and refined down through the millennia, since it is not universal among the primates. Such a system, in fact, is not the pattern of our near kin, the chimpanzees, which live in more open forests than the gorillas and at lower altitudes.

A young Englishwoman, Jane Goodall, lived among chimpanzees at the Gombe Stream Reserve near the eastern shore of Lake Tanganyika. Like Schaller, she ingratiated herself by leaving food for the animals so that she could observe them at specific places. She was soon able to recognize individuals, to know their idiosyncrasies and patterns of behavior. She found that their social arrangements were casual and certainly quite different from those of the baboons and gorillas. She noticed that young chimpanzees learned to choose the proper food and to use simple tools by imitating their elders. A youngster practiced probing into a termite colony with a plant fiber, or a twig, causing soldier termites to seize the twig and fasten themselves to it. Then the young chimpanzee withdrew the "tool" and ate the termites. Chimpanzees clumsily use sticks and stones as weapons when they come upon a large snake or a leopard, and in the noise and confusion created by these flying missiles, the leopard or snake usually seeks only to escape.

Chimpanzees often band together in small groups of two or three, then disband to join another group. Territory appears to mean much less to them than it does to baboons or gorillas. When a female becomes sexually receptive, she may offer herself to several males in succession, and there is no attempt by any one male to monopolize her. A male chimpanzee may defend his supply of food, or he may share it with a friend. A group of chimpanzees has no definite size, nor do its creatures have a lifetime membership in the club. These animals appear to have extended the idea of "family" to its ultimate limits.

The meticulous field work of researchers like DeVore, Schaller, and Goodall provides us with great masses of behavioral material which are being matched against observations made of animals in captivity or those in tightly controlled conditions in the wild. A visitor to the Jersey Wildlife Preservation Trust on the island of Jersey in the English Channel may be lucky enough to see the director, Jeremy Mallinson, reach into a welded-steel enclosure to scratch the head of a mother orangutan. As he scratches, she reaches out her fingers and touches his ankle, then puts her fingers to her nose and sniffs. This is done to remind herself of the identity of this familiar person. No amount of research in the tree-tops of Borneo would reveal this small insight into behavior.

But even more difficult to observe is the family life of a social insect. Wasps, bees, ants, and caterpillars gather together in societies that are fiercely protective of their young. Midges and other small insects swarm in thousands while the males catch and mate with females. Whirligig beetles gather in flotillas on ponds and streams. Young tent caterpillars thrive in the protection of the silken tent that they build, and enlarge, in trees.

In termite societies, the young and old work together. The termites are among the most sociable of all the social insects, their dependence on each other caused perhaps by their need to have microscopic protozoans in their digestive tracts to process the cellulose in their food. Without the protozoans, the termites starve to death. Yet they lose these tiny assistants every time they molt, since they shed the lining of the digestive tract and its contents along with the outer covering of their bodies. The newly molted termite must replenish its store of protozoans, and it does this by soliciting them from another termite that regurgitates a few drops of food, or by swallowing fresh fecal pellets.

Sharing is a central feature of termite behavior. It enforces what appears to us to be a special kind of democracy in which the physical needs of every member are provided for. It sees that food gathered by workers on foraging expeditions is distributed. It nourishes the growing young. Only reproduction is not shared. Except for the queen and her consort, all sexual drive is absent or suppressed. The two fertile termites mate repeatedly and supply all the eggs needed for the colony. The royal pair are longer lived than their offspring, but neither is indispensable. If the consort dies, a male worker matures into a sexual adult and serves the old queen. If she dies, a female worker develops in the same way to take her place.

In termite society a network of signals is transmitted constantly from creature to creature via pathways whose existence we can only suspect. Termites groom each other extensively, licking the entire surface of the body. They could easily transmit chemical messages this way, but these so-called social hormones have not been identified. The termites must have some way to coordinate the behavior of workers of all ages, of soldiers and queens and consorts. Somehow, by sharing food and protozoans and grooming, the creatures in the mound are able to make their lives and work consistent.

Social wasps, bees, and ants have evolved along separate but parallel paths. The dispersal flight is combined with the mating act, after which the males die. The pregnant females rear the initial broods themselves and from then on leave all the work to their sterile daughters. Ants transport larvae and pupae, all their efforts geared to perpetuating the heritage. That such complex behavior can be practiced by such apparently simple forms of life probably has something to do with the recognition of kinship. This would give the social unit a driving force and a homogeneity it could not quickly establish in any other way.

In an ant society, with perhaps millions of members, recognition must be impersonal and immediate: the scent of the colony itself. Any intruder must be instantly identifiable. By comparison, social behavior in a vertebrate society must be flexible enough to give each individual some potential to reproduce, and the division of labor need not be so obvious. Hierarchy replaces the caste system of workers, soldiers, queen, and consorts. In a primate society each animal must know all the others and respond to them in a suitable manner.

Looking at the teeming masses of termite and ant civilizations, it is tempting to see parallels between them and our human world. Today the numbers of men and women are so enormous, the proximity to others so immediate, the demands for energy and resources so great, and the division of labor so extreme, that the welfare of the common man seems best served through the kind of regimentation used by the social insects. Indeed, that is more or less what is happening in almost every country on earth. But in making such comparisons, because it satisfies many of our deep uncertainties about the sufferings of our fellow men, we must understand that we, too, have our ancient traditions, and that they are inherited from primate societies, not from ant or termite mounds. Among primates, it is the characteristics the individual inherited from its ancestors that are prized, and it is this individuality, not the mindless striving of the insect masses for a common objective, that makes civilization as we know it possible.

The Parasites' Path

Photographs by:
J.A.L. Cooke, 278
Oxford Scientific Films, 267, 269-273
Carl Roessler/Sea Library, 279
Edward S. Ross, 265, 274-277
P. Ward/Paul Popper, Ltd., 268
Larry West, 266
Maria Zorn, 280

The parasite moves through all levels of animal life. It is most often the final arbiter of life. It possesses innumerable devices for breaking through physical defenses, penetrating organs, feeding in intestinal tracts, attacking blood cells, tissues, creating fevers, chills, sweats, lumps, bleeding, debilitation and death.

The parasites may be lethal where their work kills the host quickly. They may work periodically so that the parasite only uses the host animal for a small part of its life span. A parasite-ridden animal, unaffected by the creatures inside it, may pass on a disease to another animal which is grievously affected by it.

Parasites are a constant test to the host animal—exploring the defenses, forcing the production of antibodies and other forms of resistance. Thus, they provide a vital role of weeding out the unfit, testing the genetic endowment, killing the sick and the old. They are manifestations of natural selection at work.

At the same time, parasite may have to contest parasite for a share of the host's resources. The host animal may be able to withstand the impact of one parasite but be killed when a second species of parasites joins the feast. Or, the host animal may have developed such resistance that it can support dozens of different species of parasite at one time.

Despite the fact that they probably kill more animals than any other single force, parasites also do much to ensure the collective health and vigor of all groups of creatures.

Parasites have widely differing effects on their hosts. Some slowly siphon off a little strength; others kill their host rapidly and feed off the carcass. The damselfly in a freshwater pond can carry a cluster of parasitic mites, each with a bloodsucking beak pushed into the host's body, for some time and still function (265). A grasshopper nymph in southern Michigan can also live for a time despite the mites on it (266). The white, rice-shaped particles on the skin of a caterpillar look like eggs but are actually cocoons of the parasitic braconid wasp. Within each cocoon the wasp maggot, feeding on the caterpillar, is transforming into a winged adult (267). Internal parasites emerge from the body of their dead host and continue to feed (268). All parasites work on precisely timed schedules designed to ensure their own development so that their survival is assured without regard for the host.

265

266

267

268

269

270

271

272

273

The cuckoo, a resident of much of Europe, lays its eggs in the nests of other bird species. The cuckoo egg here is in a hedge sparrow nest (269), the newly hatched cuckoo shoving the young hedge sparrows out of the nest (270), the cuckoo preparing to push a new hedge sparrow egg out of the nest (271), the one-week-old cuckoo being fed by the hedge sparrow mother (272), and the cuckoo at two weeks being fed by the hedge sparrow (273). Some insect mothers sting prey into paralysis but do not kill them. Then they lay their eggs on their victims; when the larvae have hatched, the hatchlings feed on the fresh meat. Here a wasp maggot feeds on an anesthetized spider (274). The braconid wasp chooses ladybird beetles as hosts for its young. After developing as an internal parasite (275), it emerges and spins a cocoon beneath the dying body of the beetle (276).

275

276

274

Destroyers

The path of the parasite through the body of its host may take almost any route. Some parasites occupy their hosts for only a brief part of their lives, others spend almost their entire life cycle at the expense of their victim. The egg laid by a two-winged tachinid fly on a tent caterpillar (277) will soon hatch into a maggot capable of entering the caterpillar and feeding as an internal parasite. The caterpillar will die. The fresh-water leech, *Piscicola*, uses suckers at both ends to hold fast to a stickleback (278) while sucking a blood meal; if the fish is small, the loss of blood may be fatal. Despite pugnacious behavior in its native coral reef, a soldierfish may be slowed down or killed by a parasitic crustacean that catches hold and takes a large blood meal (279). The larva of a wasp emerges from the skin of an anesthetized black swallowtail butterfly larva on which it has been feeding (280).

277

278

279

280

Hunters kill their victims and gain life from the death of others. Parasites achieve the same result, though they have no need for strength to humble a lion, an elephant, a tiger, or a mighty whale; they get what they want by subtle stratagem rather than by the brute power of muscle. So pervasive are the parasites that some of them live off each other in an association called hyperparasitism. Even microscopic creatures do not escape, Amoebas, for instance, live in certain parasitic, one-celled protozoans. Moreover, a host and its parasite need not be of different species. In an example of what is called sexual parasitism, the male of some species of angler fish attaches himself to the female by the rim of her open mouth when they are both young and small. He fuses his tissues with the female's and shares her circulating blood. In effect, he becomes a parasite and begins to degenerate so that soon he is nothing more than a sperm-producing appendage of the female. He no longer has any need for eyes, or the senses of smell and taste, and all are jettisoned. He only needs to know when it is time to fertilize the female's eggs, and she tells him the moment, probably through the sex hormones in her bloodstream.

There is a species of roundworm that combines both sexual parasitism and hyperparasitism. The female roundworm lives in the bladder of a rat, while the degenerated male lives inside her uterus. Another alternative is social parasitism, which is frequent among some wasps, bees, and ants. Mated decapitans ant queens in Africa investigate the nest entrances of common black ants. If the decapitans queen is not attacked by a black worker and hauled into the nest as food, she may be able to reach the black ants' brood or the black queen herself. Once settled, the intruder is not molested and is given food. Gradually she takes on the odor of the black colony and is free to wander. Her goal is to get on the back of the black queen, and once there, she earns her scientific name—decapitans—by biting off the head of the black queen. Now she is the only surviving female in the colony able to lay eggs, and she produces decapitans eggs, which the black workers rear. New females will emigrate to find fresh centers of free labor. But the transformation of an organism into a true parasite involves a series of progressions over many generations. At first, it is merely convenient for the parasitic creature to be partially dependent on the resources of its host. Then, as evolution proceeds, new generations of parasites become completely dependent on their hosts and begin to speciate into separate groups of creatures capable of living off animals with habits similar to that of the original hosts.

The Parasites' Path

The development of louse-flies shows the depth and scope of this long process. About one hundred species of these flies are spread around the world; all are equipped with clawed, powerful legs and flattened, leathery-looking bodies, and include both winged and wingless creatures. Their parasitic history is ancient. One species parasitizes large woodland birds; another species lives off small perching birds; a third species concentrates on birds of the moorlands. When frigate birds are caught up in tropical storms and driven across the Atlantic from the Caribbean to Europe, they bring with them some of these parasitic flies, which cling to their feathers. When a bald eagle robs an osprey of its fish meal, the brief contact is apparently enough to pass on to the osprey some of the parasitic flies attached to the eagles.

These flies probably originated as parasites in the nests of birds, and in their most highly evolved form they have become wingless. This is no disadvantage, although the flies are linked to swallows, swifts, and martins, which spend most of their lives on the wing and the parasites cannot pursue them. Instead, the flies lie in wait at the nesting colonies of the birds. The pupae winter in the nests of their victims, and when the birds return in the spring, the adult flies are waiting for them. The parasites are so successful that most birds of prey—the owls and falcons, hawks and eagles—are also usually infested, having picked up the parasites from birds they have killed.

Less highly evolved members of this family of parasitic flies are those which have concentrated on mammals. There are both winged and flightless species among these flies, too. Those without wings, called keds, have speciated into forms that parasitize horses and mules, cattle and camels, dogs and ostriches, deer and sheep, chamois and kangaroos, wallabies and lemurs, among others. One of these flies, which parasitizes deer, breaks off its wings after it has mated and found a desirable host, thus making a total commitment to a single animal.

Whether a parasite lives safely or dangerously depends on how well it is adapted to its host. The parasitic pearlfish, which backs its transparent body tail-first into the vent of a sea cucumber, has found nearly perfect protection; it is safe from its enemies and it can feed on parts of the sea cucumber's internal organs. But another kind of parasitic pearlfish is not so secure, for it lives in an oyster which can trap the fish and cover it permanently with mother-of-pearl.

Parasites which damage their hosts (pathogens) cause a multitude of diseases. This group, almost all of them invisible to man's unaided eye, include bacteria, spirochetes, funguses, viruses and rickettsia. The work of pathogens is so widespread that practically no part of any animal's body is immune to their attentions. And there is hardly any limit to their capacity to weaken, maim, transform or kill their hosts. The sum of their work synthesizes the geography of almost all disease.

Parasites that cause traumatic damage are usually visible. Tapeworms and hookworms cause death, or extreme sickness, when their larvae or adult forms migrate through the bodies of their hosts. They puncture the walls of intestines, rupture blood vessels, and damage vital organs as they move onward, either to feed or to begin another state in their development. Some parasitic worms are so big that when they leave the host body, large circular holes remain which look as though they have been drilled by a machine.

Yet another parasite group, called proliferative, exemplified by the amoebae, affect their hosts only slightly. Amoebic colitis may debilitate the host, but it does not damage the tissues. Others in this group upset the production of red corpuscles. Certain worms cause the tissue of the host to grow. This may seriously affect the functioning of either tissues or organs. However, while the proliferative parasites do not kill, the upset caused by their work may be serious enough to let in other parasites, particularly bacteria, which subsequently cripple or kill the host animal.

Generally, parasites get their sustenance as easily and as directly as possible. Microscopic parasites teem in the digestive tract of animals where they feed on recently ingested food. The host animal can usually sustain this without serious impact. But when the parasite is large, such as a big tapeworm, it may eat so much of the host's food that the creature starves and dies. Some hookworms also suck blood from the tiny blood capillaries in the walls of the intestines. If scores, or hundreds, of hookworms are present, they may cause serious anemia. Giant intestinal roundworms, and some tapeworms, which may grow 25 feet in length, so completely fill the intestinal tract that they block the flow of vital materials such as bile and urine, and other materials vital to the functioning of organs.

The lytic parasites work in yet another way, digesting cells and tissues by releasing juices which cause fermentation. Other parasites release metabolic products that cause such irritation that they literally poison the creature in which they are living. Allergenic parasites sensitize the host by releasing proteins and other substances. The host creature suffers chills and fever which may kill. In man, malaria is a typical manifestation of allergy.

As parasites make their manifold ways through the tissues and organs of their hosts, the afflicted creatures fight back, strengthening their natural resistance to the enemy inside. Many African animals have the trypanosomes of sleeping sickness in their bloodstreams. The trypanosomes may kill a man but they have no effect on the lion. The common rock dove, the familiar city pigeon, is host to about seventy species of animals and plants which live on, or in, its body. It is attacked by tapeworms and flukes, protozoans and fungi, roundworms and lice, bacteria and viruses. It may be infested with ticks, mites, one species of fly, and bugs. As many as thirty flies may travel in its feathers at one time. A badly infested pigeon may be burdened with literally thousands of bird lice. Its intestines may contain up to one thousand tapeworms.

The parasite provides an unfolding picture of the infinite diversity of life itself. No concept is too bizarre for the parasite to realize. A crustacean parasite, *Succulina*, attacks a crustacean, the crab. It becomes a part of the crab's body, with segments of the parasite's body projecting from the abdomen of the crab. It grows, somewhat like a cancer, sending rootlike growths to every part of the crab's body. But while it is feeding on the crab's blood, and growing, it does not effect the vital organs of its host.

However, in an odd twist of the parasitic process, *Succulina* does affect the sex glands of the male crab, and makes it gradually more and more female. The large claw grows smaller. The swimmerets swell and begin to look like the swimmerets of a female crab. The parasite only affects the male crab because its early sustenance comes from a fatty substance that is more abundant in the female crab's blood and is needed to produce eggs. The parasite eats this and prevents the eggs from developing. The ovary degenerates and the female becomes sterile. But once inside the body of the male crab, the parasite does not get enough fat and it must make heavier and heavier demands on the crab until the process of physical degeneration and sexual transformation is underway.

In some exceptional cases, this parasite actually creates hermaphroditic crabs as the crab, in what appears to human eyes to be a desperate effort to save its life, becomes bisexual.

Parasites influence practically every form of life on earth, and probably do more to control animal populations than any other single force. They certainly help to determine the numbers of lions in the African savannah, and to influence the success or failure of the massive breeding of antelope on the high African plains. They affect the biology of fish in the deepest parts of the sea and of the highest sailing eagle or condor. Parasites reach close to the elephant's heart, eating away at its tissues. Parasites live in the gullets of deer, in the kidneys of pigs, in the livers of tigers. A tiny South American catfish lives in the gill cavities of a larger catfish, where it feeds on its host. Flatworms get into digestive tracts and blood vessels. Tapeworms, threadworms, and roundworms easily penetrate the bodies of many animals. Parasites live in the eggs of other creatures and are released when the eggs hatch. Some fish lice fix themselves to the bodies of fish and are able to move around by adjusting the grip of their circular suckers. Others bury their heads as a tick might into the flesh of fish.

The interplay between parasites and hosts extends into the insect world. Almost half of all known insects are dependent on each other, in one way or another. The tomato sphinx moth caterpillar, bumping its green body over leaves in search of sustenance, is equipped with a long and lethal-looking spine up its back, but this does not deter the parasitic larvae. Their cocoons cover the caterpillar's body and the larvae will eventually eat it alive. An aphid, its mouth parts buried in the stem of a plant, may sense the black wasp landing on its body and desperately extrude droplets of a repellent substance, but the defense tactic does not work. The wasp carefully inserts an egg into the abdomen of the aphid. The most lethal parasites kill their hosts quickly, but others work without ever revealing their presence, so that both creatures coexist, the one living off the other, until the compromise is no longer necessary. A caterpillar's body may become so filled with developing fly larvae that it collapses and dies just as the larvae are ready to mature.

The work of parasites is so pervasive that the host species have been forced to develop numerous ways to repel them. The great rain forest moth larva grows long poisonous spines that more than double its body size. The fat-bodied plume moth caterpillar is covered with fine hairs that hold droplets of glue that trap any parasitic wasp seeking to lay eggs on the caterpillar's body.

Like the rest of life, parasitism is infinitely varied. A parasite may spend all its time on its host, or very little time at all. A sea lamprey, for example, is not parasitic when it is young, but as an adult it fastens by its sucker mouth to the body of a fish. The lamprey may use its rasping tongue and extremely powerful saliva to convert the flesh of the living creature into a thick puree, or it may just suck its blood. Once attached, the sea lamprey remains motionless while it feeds. The fish usually dies from loss of blood, or from the deep wounds inflicted by the lamprey's tongue. The sea lamprey has a smaller oceanic relative known as the slime eel, one of the few other parasitic vertebrates. Sometimes this eel will cut a passage into the body of a dying fish and eat the creature from the inside. The slime eels are so omnipresent and so determined to get a meal that harpooned whales are often penetrated before they can be hauled from the water.

Most bloodsucking leeches, relatives of earthworms and clamworms, could probably survive by eating carrion or by hunting insects, snails, and worms. But they prefer the blood meal. They are wide-ranging in their choice of victims, attaching themselves to most cold-blooded creatures, mammals, and birds. The various leeches are prefectly adapted to their parasitic tasks. A sucker at each end of their bodies gives them the capacity to climb anything, to travel with looping movements, and to hang on tightly to any support. They can store several times their own weight in blood, which their saliva prevents from clotting. The leech gets rid of most of the water from its blood meal and then stores as much undiluted blood as possible. The horse leech fastened to a cow that has wandered into a pond for a drink can stretch to more than twelve inches before dropping off to digest the blood meal.

Like most other parasites, different species of leeches concentrate on specific victims. The common brook leech prefers fresh-water mollusks; the flat-bodied leech lives off turtles while other leeches seek rays or skates; one marine leech parasitizes oysters. But not all leeches are dependent on water to reach their victims. Land leeches teem in Southeast Asia and in Australia; they lurk in the damp undergrowth and wait for a chance to attach themselves to almost any living thing that passes. They can clip through the eyelet beside a bootlace, penetrate a man's sock, and within an hour increase from the size of a toothpick to that of a small cigar.

The bite of a leech is scarcely detectable, and never as painful as an insect bite. The wound is made with three lancets situated inside the leech's mouth. They produce a Y-shaped cut with clean edges. The leech has a good chance to start sucking blood before its victim is aware of what is happening. The cut is so painless and the extraction of blood so efficient that medicinal leeches are still raised in carp and duck ponds in many parts of the world. Because a leech will drop off its victim once it has had a full meal, or can be driven off by sprinkling it with salt, one leech can be used repeatedly. Its anticoagulant saliva is so strong that it diffuses into nearby blood vessels and liquefies clots. A leech can banish the discoloration of a black eye by feeding around the damaged area. But these parasites are used by uneducated people with misplaced confidence in their capacity to cure a variety of diseases.

The leech can become the agent of infection by exposing the host to other parasites. About 80 per cent of all the infectious diseases of man are spread by a limited number of biting insects, mites, and ticks. A man watching the approach of a malarial mosquito has no idea whether the creature contains the protozoans that would give him the disease. He cannot tell whether the deer fly that has just bitten him is infected with tularemia. It is small comfort for him to know that the males of mosquitoes, horseflies, ticks, and chiggers stick to a diet of plant juices, and that only the females bite.

Sand wasps capture live insects to feed their carnivorous young. They are technically referred to as "parasitoids," since they exploit other animals only at certain times, in this instance to nourish growing larvae. The cicada-killer is a large wasp that wrestles its powerful prey into submission and then, with great effort, flies with its burden before dragging it into its burrow. Some smaller wasps stock their nests with aphids, and the mud dauber wasp fills her cells with spiders. Another wasp, the tarantula hawk that frequents the American tropics, stings the huge and hairy spider before bringing its anesthetized victim to its young. Each hunting wasp injects just enough venom to tranquilize the victim, as it were, without killing it. The venom must be adjusted exactly so that the creature is still alive, but helpless, when the wasp's egg hatches out and the larva can begin eating. Ideally, larvae eat first those parts of the creatuare that are not vital to its life, leaving until the last those organs that are essential to it. At this point, the parasitic larvae must quickly become scavengers to use up the rest of their food. But to find a victim, anesthetize it, and then carry it to the young suggests the possibility of a much easier method—that of carrying the egg to the victim and laying it on the host directly. The small, bristly tachinid flies and the wasplike ichneumon flies both hunt for caterpillars, and when they find the right species, the mother alights delicately on her victim to cement her egg to its back or jab the egg into its body. The egg will hatch in, or on, the caterpillar, and the larval parasite will begin eating. At first, the caterpillar is merely a host, and much bigger than the parasite growing on its body, but eventually it dies as the appetite of the parasite becomes overwhelming. With its host dead, the larva pupates and transforms into an adult, its parasitoid period ended.

To survive, the parasite must have a successful host, and it must continue to colonize young, healthy hosts without killing them. This is the great check on the multiplication of parasites. Clearly, it is difficult to find a host animal, and once one is found, the parasite behavior must match the host's habits.

Parasitic worms follow strange paths to their final homes. Adult tapeworms usually live in the small intestine of meat-eating animals, where they hold themselves in place with hooks or suckers or both while producing a series of young, which appear to be additional segments in the tape. The whole worm, except the first attached segment, swims about inside the liquid that fills the small intestine; this prevents its body from interfering with the process of digestion. Each segment is an individual, which grows as it absorbs food, and develops male organs, then female organs, and finally embryos. Sperm from younger segments fertilize the eggs in older segments. The embryos develop to quite a size before the segments carrying them break away from the free end of the tapeworm and exit from the host in its feces. Now the embryos must find their way into an intermediate host, which may be a pig or a bear for the pork tapeworm, or a cow for the beef tapeworm, or a fish for the fish tapeworm. In its final host, the pork tapeworm may grow to twenty feet; the beef tapeworm, thirty feet, and the fish tapeworm, sixty feet. Whether pig, cow, or fish, the intermediate host can pass on the parasite to a human being who eats the animal's flesh. Transfer is usually prevented by thoroughly cooking the meat before eating.

Children occasionally swallow infected fleas and so become the final hosts of dog tapeworms, which grow to a foot or more before the existence of the worm is suspected.

In New Zealand a tiny tapeworm, which infects dogs when they eat infected sheep flesh, is very dangerous to humans. The tapeworm eggs, contained in dog droppings, can get into human food, and once inside the body they form brood capsules that often become as large as an orange. If these capsules, called hydatid cysts, develop in the membrane covering the brain and cannot be removed surgically, they cause death. New Zealand law requires that every dog be purged twice a year and its wastes inspected for tapeworms by a veterinarian.

All flukes are parasites, and a few of them use only one host. Some fasten themselves like leeches to the gills or skin of fish. Others hide in nasal passages, in mouths, or in the bladders of amphibians and reptiles. The majority of flukes, however, live in two or three intermediate hosts before they reach the animal in which they can mature. Their bodies change for each host, and their transformations are the most complex of any animal. In all this activity there is a general pattern; the final host is almost always a vertebrate in whose feces the fluke eggs will reach water.

The first intermediate host is usually a mollusk, which is never eaten by the final host. The larval flukes living in the mollusk produce new creatures, which disperse in the water. Flukes from fresh-water snails give birth to fork-tailed larvae that swim in the ponds and streams of Africa. They can burrow into the wet skin of a human being and swiftly enter the bloodstream to become blood flukes. They may stay inside the human body for a lifetime, dwelling in the blood vessels in the wall of the large intestine. There, a pair of them will mate, the male curling his thin, leaf-like body around his longer and more cylindrical partner. Whenever she is ready, the female fluke perforates the intestinal wall and deposits a batch of fertile eggs where they will be evacuated with the feces. The constant perforations of the intestinal wall eventually may make the host anemic from loss of blood.

The routes by which parasites find their next hosts are infinitely varied. Some worms must be eaten in their larval stages if they are to develop. The single-cell parasites that cause malaria reach mosquitoes in a blood meal. They mature in the insect and produce a large number of young, which live in the mosquito's salivary glands. But the protozoan must reach more mosquitoes and does so by entering the bloodstream of an intermediate host, usually a mammal. There, it is carried to the liver, the bone marrow, and other parts of the body. It reproduces asexually by invading red blood cells, digesting their contents, and dividing into a dozen or so tiny new individuals, which escape from the ruined red cell and spread the infection to other cells.

Through all this, the parasite multiplies according to a rhythmic cycle. All the escapes and reinfections occur on the same day or during the same night, which is why the infected man may feel healthy enough to live normally until an explosive escape of parasites and wastes from the red cells makes him feverish and weak. He wants only to lie down quietly—and this is precisely what helps the parasite. The fever makes its victim's body hot, thus helping mosquitoes to find him. He is weak and less likely to kill the mosquitoes that are continuing to spread the infection. The parasite times its releases for the daylight hours if the local mosquitoes bite by day. But if the patient suffering from malaria moves to a district where the mosquitoes bite by night, the parasite changes its schedule in just a few days.

Many of the diseases caused by parasites seem to appear in intermediate hosts that are expendable. The well-adapted parasite usually gets what it needs from its final host without making its presence too obvious, and adjusts its actions according to the amount of its food supply. Ideally, it should not become so numerous or voracious that it kills its host, and it should not interfere too much with the reproductive capacities of the victim. It should prey on the aged and the young to avoid seriously depleting the populations of its hosts.

In the world of the parasites there is a sense of stability and continuity that have rolled on like slow-breaking waves for more than a billion years. Yet this is illusory. The adaptability and sensitivity of all organisms remain the fundamental forces for change and development. All members of every species work incessantly to expand their share of the limited space and energy the world can give them. But as fast as one creature achieves a breakthrough, another counters with a new defense, so that while animal communities appear to be stable, they are, in fact, diverse and ever-changing.

Guests and Hosts

Photographs by:
Stan & Kay Breeden, 302
Bruce Coleman, Inc., 304
Arthur Christiansen, 290
J. & C. Church/Sea Library, 284
Francisco Erize, 291
Keith Gillett, 289
Grant Heilman, 281
R. Mariscal/Bruce Coleman, Inc., 282
Karl H. Maslowski, 293
Oxford Scientific Films, 297
F. Park/ZEFA, 303
Carl Roessler/Sea Library, 283, 286, 287
Edward S. Ross, 298, 306
Tui De Roy, 294
Leonard Lee Rue III/Bruce Coleman,
Inc., 295
F. Sauer/ZEFA, 300
Caulion Singletary, 301
Ron and Valerie Taylor, 285
Ron and Valerie Taylor/Sea Library, 288
Simon Trevor/Bruce Coleman, Inc., 296
P. Ward/Paul Popper, Ltd., 305
Peter Ward/Ecology Pictures, 292
Larry West, 299

In southern Florida, burrowing owls share the excavations of gopher turtles. On the prairies, they share the dugouts of prairie dogs. The dogs and turtles that dig the shelters eat an occasional egg or chick. But this is not serious enough to deter the habit of the owls.

Many animals have developed extraordinary habits of cooperation—or so it seems to human eyes—that benefit both parties. Animals with these habits are called commensals—from the Greek *companions at the table*. For instance, a certain species of beetle lives in ant nests by either massaging the ants so that they will disgorge food, or by raiding their storage larders. The beetle lives without disrupting, or even affecting, the ant colony.

Cooperation may be symbiotic. Then, animal or plant associations make life possible for both. A fungus and an alga combine to make a lichen. Neither can function without the other. Many tropical creatures hitchhike with others. The suckerfish, a tropical and temperate ocean-ranging creature which is not a parasite, attaches itself by a suction cup to larger fish or to a turtle and saves the energy of swimming itself. A botfly lays a fertilized egg on a mosquito, which it has captured, then releases the mosquito unharmed. The egg hatches when the mosquito is taking her blood meal, and the botfly maggot drops off onto the animal being used by the mosquito, bores into the skin, and matures there.

In addition to this sequential type of cooperation, there are instances where several animals work together. A certain anemone needs the contact of a smaller fish to activate its tentacles. The smaller fish then shares the prey caught by the anemone.

The scope of inherited and learned behavior among animals is truly vast.

Some sea anemones, fixed to tropical reefs, have developed complicated commensal relationships. These are a form of cooperative behavior, or the appearance of cooperation, for one or both of the creatures to secure a living. Normally, the anemone, such as the American *Bundodosoma,* captures and engulfs a striped "high hat" when it brushes against the outspread tentacles (281). Such straightforward hunter-hunted activities are the most common. But some Pacific Ocean anemones, such as the *Calliactis,* will not fully expand their tentacles until one particular species of small damselfish, brushes against them (282). The damselfish, however, is not the victim. It is only a trigger to warn the anemone that hunting may be possible. The damselfish is actually only a decoy and is never caught. Instead, another fish, darting at the damselfish, is itself the victim.

281

282

283

284

285

286

Small creatures in the coral-reef community, such as shrimp, are hunted by large fishes not as food but because the shrimp can remove parasites from a fish's body. Often the big fishes line up to get a chance to be cleaned off. Hawkfishes of the Indo-Pacific wait for large, parasite-infested fish (283). A cleaner shrimp with conspicuous antennae works over the head of a moray eel (284). A small, distinctively marked wrasse also cleans parasites from a moray eel (285). A juvenile hogfish will clean a much larger tang (286). A small wrasse will clean turtles as well as larger fish (287). The tentacles of a jellyfish are poisonous to any creature without a shell; the tropical leatherjacket can, however, swim unharmed among the trailing tentacles and so gets protection while waiting for its own prey (288). Goose barnacles attach themselves to surface-floating purple snails (289).

287

288

289

Birds as Cleaners

The interaction of one creature's life with another is worked out in a multitude of ways. Parasites infest the bodies of their hosts. But cleaner birds follow the hosts, the larger African animals, and eat the parasitic insects. A gazelle will hold still to let a tickbird remove a parasite from its ear (293). A young zebra learns early to tolerate the vigorous searching of its body by oxpeckers (292). The rhinoceros is frequently decorated with several cleaner birds working over its hide (290). Cape buffalo are not only cleaned by cattle egrets but the wary birds, standing high, give early warning of danger (291). Land tortoises as well as large mammals are served by cleaner birds (294). Hippopotamuses, half submerged in a water hole, are cleaned by cattle egrets, which also warn of danger (296). Cleaner birds eat not only the parasites on African elephants (295) but also the insects that the elephants stir up.

290

291

292

293

294

296

295

169

Insects fertilize pants by carrying pollen. Pollination and the plant's continued life are often essential to the insect. One species of orchid produces a blossom so exactly resembling the female bee that the male is deceived into copulating with the flower and carrying its pollen (297). The honeybee is often completely dusted with yellow pollen when it has drunk the nectar from a flower (298). The long-horned borer wasp drinks from the flower of the prickly pear cactus in the American Southwest and at the same time disseminates the pollen (299). Two moths simultaneously pollinate the common blood drop plant (300). These butterflies—gulf fritillaries—help pollinate the "blazing star" flowers (301). Even small mammals, such as this Southeast Asian lemur, spread pollen (302), as do various nectar-eating birds such as the hummingbird and the honey-eater (303, 304).

297

298

299

300

301

302

303

304

A long-term relationship may be formed between two animals that is neither parasitic nor completely commensal, yet they may become dependent upon each other. Ants use their antennae to stroke plant lice, or aphids, which respond by extruding "honeydew" out of tubes projecting from their abdomens. The honeydew is a waste solution of sugar and water remaining after the aphids have extracted protein from the plant sap. Some species of ants have "domesticated" the aphids, taking them into their chambers where the aphids are fed on roots and are "milked" by the ants. Here, ants milk two kinds of plant lice on the branches of trees (305-306). The life histories of ants contain many similar examples of "exploiting" other creatures, of growing underground crops of fungus for food, and even of "enslaving" other colonies by penetrating their nests and breeding their own species.

305

306

Concealed in the galleries of almost any ant nest in the world are tiny beetles, inconspicuous creatures that act like ants, smell like ants, and often resemble ants in size, color, and bodily proportions. Sometimes one of these beetles approaches an ant and offers it a massage. It strokes the ant with its antennae or mouth parts, or with its forelegs. If the host ant challenges the humble beetle, it quickly offers the ant a regurgitated droplet of food. If the food is accepted, the ant is not likely to attack the donor.

The beetle probably obtained the food by massaging another ant, or by stealing the food from the storage rooms where the ants keep their provisions. As long as the nest larder is full, the beetle is a comparatively harmless lodger that pays for its room and board by massaging ants and offering them food. In these happy circumstances, the beetle is called a commensal, which means "companion at table." But if the ants grow short of food, the uninvited beetle guest becomes a parasite, or an inquiline, which means it has been reduced to the status of an unwanted tenant.

The Romans might have called the beetle's behavior in times of plenty *quid pro quo*—"something for something." A frontier storekeeper would have called it barter. In government circles it would be known as trade or exchange. Whatever it is called, mankind has no monopoly on swapping, which underpins much of the behavior of animals everywhere.

In 1879, a German botanist named Heinrich De Bary coined the word *symbiosis* to describe how different species of animals or plants associate with one another. Similarly, many different animal and plant forms reap positive benefits from living together. The fungus and the alga, for example, together make up the lichen. Scientists now subdivide such relationships according to how much one species can spare for the other to take, and on whether the association is optional or obligatory.

Cooperative behavior is especially varied in the rain forests of the tropics, where challenges to survival are more likely to come from the neighbors than from the environment. There, living together may be as simple as the arrangement of the oriolelike birds named oropendolas, which build their pendant nests close to colonies of wasps. The wasps do not bother the oropendolas, though the insects will viciously sting most other intruders. As a result of their close association with the wasps, the oropendolas are protected from their many predators and can raise their nestlings in relative safety.

In southern Florida, burrowing owls share the excavations of gopher turtles, and west of the Mississippi they share the dugouts of prairie dogs on the plains. The dogs and turtles that dig the shelters in the first place eat an occasional egg or chick, but apparently the price is not too high. The season of the year may determine whether an animal benefiting from living with another species is doing its host any harm. A young lion may not be inconvenienced by tapeworms in his intestines until he begins to starve and the worms are competing for scarce food, or if he becomes too old to hunt efficiently. The tapeworm is even more delicately balanced. It cannot transfer to another host when life gets tough; it survives only by taking as little as possible of the host's strength.

One of the rove beetles that commonly occupy ant nests goes beyond stealing food and placating its hosts with massages. By a specialized procedure it becomes a guest in the brood chambers where the ants keep their young. To gain entry into the ant nest the rove beetle emits a quick-acting tranquilizer from the tip of its abdomen, and when the nest guards sample it, they lose their aggressiveness. The beetle then turns to let the tranquilized ants lick "adoption" glands at the sides of its abdomen. These glands produce a hypnotic drug which, changing the guard ants' behavior once more, causes them to welcome the beetle with more than ordinary warmth. Indeed, the guards pick up the insect and carry it directly into the brood chamber. The helpful beetle rolls up tightly to assist the ants in transporting it. Once inside the brood chamber, the intruder is free to wander about and eat the ant larvae. There it will probably meet a beetle of the opposite sex, mate, lay eggs, and produce young.

The larval beetles creep about the brood chamber secreting minute droplets of a substance that attracts the ants. They accept the proffered secretion and give the beetle larvae regurgitated food in exchange. The young aliens supplement this diet by eating the smallest of the ant larvae. But in this most perfect of all parasitical worlds there is one flaw. The beetles must somehow control their own population; their larvae are cannibals and feed on one another before they eat the ant larvae.

Some ants that are extremely aggressive and are excellent fighters swarm from their nests at intervals to attack other colonies. They kill every worker, male, or queen they can find, but carry home the juveniles, both larvae and pupae. When these creatures mature, they swell the ranks of the worker caste within the colony. At least one species, the Amazon ants of Europe and America, cannot function without these outside workers. The Amazons' great sickle-shaped jaws make them formidable fighters and enable them to carry home slaves, but the ants do not have the body structure or the inherited programming that would allow them to hunt their own food or tend their own young. When they are deprived of their slaves, they starve to death and ignore their own offspring. Their narrow specialization makes them utterly dependent on other ants for survival.

Most commensal creatures are harmless hangers-on that accept shelter, transportation from one feeding area to another, or wastes that nourish them. Few of them come anywhere near the parasitism or predation of the rove beetle. But they all must be able to adjust to changing conditions. The tuatara lizard of New Zealand, a two-foot-long reptile, has survived relatively unchanged for some two hundred million years despite the enormous differences that have occurred in its environment. In its heyday during the Triassic period, when dinosaurs roamed the earth, there were no birds. Yet when shearwaters appeared about thirty million years ago, the lizards began to share nest burrows with the long-winged sea birds. The lizards can dig their own burrows if the soil is loose enough. But they now prefer shearwater tunnels. The lizards, which can be sexually active at one hundred years of age, keep the shearwater passages open during the nonbreeding year while the birds are at sea. When the shearwaters return to mate, they need only clean up the nest cavity at the end of the tunnel. One shearwater remains on duty at the nest while its mate goes night-fishing for food, so the eggs and chicks are always protected. When the young shearwaters are finally deserted shortly before they are ready to leave the burrow themselves, they are too big for the lizards to eat.

Although the tuataras are not completely dependent on the birds for shelter, they do have another link with the shearwaters: The birds' excrement fertilizes the island slopes where shearwaters breed. The roots of the evergreen shrubs that grow there prevent erosion following the frequent rains, and the shrubs' shade stops the soil from baking into a cake so hard that neither tuatara nor shearwater could dig into it. If the shearwaters fail to nest, the lizards disappear. These living fossils, which have seen so many changes, cannot cope with such a relatively minor difference in their island community. In an indirect way, the reptiles have become dependent on their commensal existence with the shearwaters.

The relationships of guests and hosts are as varied as the species that share living arrangements. In the coastal waters of the North Atlantic, skeleton shrimp live as guests on some sea stars. The shrimp creep over the sea stars searching for tiny particles of food, literally grooming the stars without in any way disturbing them. The sea star could use its pincerlike organs to catch the shrimp, but it does not. The shrimp are allowed to move from one sea star to another so that they need never touch bottom. Since the shrimp have no swimming stage, their sea-star hosts provide them with perfect platforms from which to launch their young.

Little pea crabs can run and swim freely, yet they choose to live in the mantle cavities of sedentary bivalves or in the tubes that worms make and maintain in mud flats. Agile oyster crabs stay with their oyster hosts through thick and thin and sometimes reach the dining table intact. Mussel crabs live inside the shells of healthy mussels, scallops, and pen shells, where they find both food and refuge. Parchment worm crabs mature, mate, and hatch their eggs in the U-shaped tubes of parchment worms. Lugworm crabs live in lugworm tunnels. All these round-bodied crabs have escape mechanisms. If a storm tosses the hosts ashore to their death, the crabs usually survive, for they burrow into the bottom sediment at the first sign of danger.

To survive as guests, the commensals should have a certain amount of flexibility so that they may have the option of choosing new hosts from another species if that becomes necessary. To measure the adaptability of some commensals, polychaete worms have been conditioned in a laboratory environment to take hosts that previously were unfamiliar or unattractive to them. The polychaete normally associates with at least nine species of animals, including certain limpets, sea stars, and sea cucumbers. When the worms were removed from a familiar sea cucumber and placed with an unfamiliar sea star, the creatures changed their host preference. Not only did fewer worms choose their original hosts, but a significant number of them preferred the new partner to the old one.

The games between guest and host go on endlessly. Some minute snails cling to particular sea anemones; other sea anemones hold on to mussel shells. Tiny bivalves stick to the rims of the unattached tubes that sandworms construct. Small sea cucumbers grasp the rough skin of deep-water angler fish, and a small fish named the fierasfer slips tail-first into the breathing opening of a big sea cucumber. The sea cucumber does not want to accept its uninvited guest. It closes its breathing opening and ceases to change the water that brings oxygen into its respiratory system. Then it waits. But the fish waits too. The moment the cucumber breathes and opens up its system, the fierasfer backs into its refuge and the sea cucumber is stuck with its inconvenient guest.

Fresh-water bivalves use adult fish as custodians of their young. Pocketbook mussels, so named because of their bulging, rounded shells, retain their hatched young in capacious gills until the tiny creatures grow a pair of shell valves. When small fish are swimming nearby, the mother expels her young, each one about the size of a poppy seed. A minnow which tries to swallow such a tasty morsel will find instead that it has acquired a guest. The tiny shellfish snaps its valves shut around a gill filament, and there it stays while the fish carries it about and keeps it supplied with fresh water. Once the mussel is mature enough to survive on its own, it lets go and, swept free by the current created by the breathing fish, the mussel settles into the mud.

The bitterling, a central European minnow, and the fresh-water bivalve that hosts the bitterling's eggs and hatchlings are partners in a complex courtship procedure. The male bitterling apparently chooses a bivalve of the proper size and then establishes his territory around it. He will move this territory if the shellfish moves, and he will not tolerate any other fish of his own kind or size in the area. He will expel even a female of his own species unless she is submissive and shows a readiness to follow him, which tells him she has ripe eggs to deposit.

The male quivers before such an available female and leads her to the bivalve. By moving his own body, he induces her to align herself parallel to the shellfish and just above it. At this point, she extends a slender, slack tube about half as long as her three-inch body, and drops one or two eggs into it. The eggs settle at the tip of the tube but do not pass through it. Then, in an action so swift that the human eye cannot follow it, she swoops down on the bivalve, thrusts the tip of her tube into the breathing opening of the shellfish, and from her muscular bladder emits liquid that dilates the tube and fires the eggs into the bivalve. The female bitterling moves aside and reloads, as it were, while the male hovers over the shellfish and floods his milt into the mollusk's mantle cavity. The eggs will hatch in about three weeks in this sanctuary, and the tiny fish will swim out against the incoming current.

The bitterlings are catholic in their choice of bivalves; when they were introduced into the waters of New York State, they had no trouble electing North American bivalves to care for their eggs. Occasionally a shellfish may clamp shut his shell and cut off the tip of the female's egg tube, but she is able to grow a new tip.

Fish are not the only creatures to use foster parents to care for their young. Birds belonging to at least three families lay their eggs in nests built by other birds. A female slips into the unattended nest, lays an egg, and departs. She may have waited until the nest contained at least two eggs, showing her that the builder had made a commitment there. Her youngster may throw out all the hatchlings of the nest-builder when it is strong enough, or it may live harmoniously with them and be reared by its foster mother. The best known of these exploiting birds is the European cuckoo, which has given the English language the word *cuckold* to describe a male whose mate raises the offspring of another male. The unrelated African honey-guides and American cowbirds also use the nests of other species for egg-laying. Sparrows and warblers, which are often the unwilling hosts, have learned to fight back. When they find a cowbird egg in the nest, they build another nest floor over the offending egg and lay a new clutch. The same cowbird, or another one, may return again, and the process of laying and rebuilding may be repeated several times in a single summer.

Man presented the cowbird with a new host species. The charming little Kirtland's warbler never devised protective behavior against cowbirds, since it nested among the jack pines of northern Michigan and wintered in the Bahamas, where there were no cowbirds. But when pioneers felled forests, brought in cattle, and cultivated grain, they enabled the cowbirds to extend their territories from the west to reach into the Kirtlands' limited breeding territory. Recent studies show that 66 per cent of all Kirtland's warbler nests are now invaded by cowbirds, and more than a third of all the warbler fledglings in these nests are lost.

When man does not interfere, the relationship between guest and host bird is usually so stabilized that no particular harm is suffered by the host. In fact, the relationship may actually help the host creatures. Oropendolas share their nests and food with cowbirds, and in such nests death by flesh-eating fly maggots is extremely rare. The young cowbirds catch the flies that come to deposit their larvae, doing this so expertly that none reach their nest mates. But we do not know why the cowbird should be so wary of the fly maggot while the young oropendola seems insensitive to the danger.

The free ride is used by countless species for short or long periods. Suckerfish that live in tropical and temperate seas attach themselves to larger fish, to sea turtles, and sometimes to fishermen's boats. In ancient times, Greek fishermen called them "shipholders," claiming that the attached fish held back a sailing boat in a good breeze. The creature uses a modified dorsal fin on top of its head as a suction cup. Recent research suggests that the suckerfish eats small parasitic crustaceans and is actually doing its host a favor. If a suckerfish wishes to detach itself, it must swim faster than its host to reduce the hold of the remarkable oval suction cup. Hitchhiking is a favorite form of travel for many insects. Some insects board female grasshoppers and wait until they lay their eggs before dropping off, knowing now where the concealed store of food is hidden. Some beetle larvae hitchhike on the backs of adult bees and wasps, which take them directly into the nests, where they can begin eating. In the American tropics, one of the botflies uses a mosquito to help it reproduce. It seeks out and captures a female mosquito, lays a fertilized egg on the underside of her abdomen, then releases her unharmed. That night, when the mosquito finds a blood meal from a mammal, the botfly egg quickly hatches, then the maggot drops free and burrows into the mammal's skin to mature there. Since the adult botfly has no way of knowing what kind of host its egg will be carried to, the maggot that hatches out must be able to cope with a variety of situations. And because most wild animals have no way of getting rid of the creature that came to them as a hitchhiking egg, the system is a success. One wingless fly, known to beekeepers as the bee louse, clings to the body of a worker honeybee as it goes about its chores. For a long time the fly was thought to be a parasite that sucked blood from the bee, but a closer study has revealed that the fly is merely commensal. It creeps along the bee's head and tickles its mouth parts until the bee regurgitates a droplet of food. Inside the hive, the tiny fly leaves its host just long enough to mate and for the female to deposit her eggs on the combs. The fly larvae sip honey from open cells and nibble pollen grains, but they have no appreciable impact on the hive. As each larva pupates and becomes a mature fly, it boards a bee and begins its flying life. It does not matter when the bees swarm; there will always be bee lice traveling with the workers, and the new queen will inherit the larval bee lice along with the old hive. Every honeybee household has these guests.

Often two creatures associate for mutual benefit without either's actually needing the other. In Africa, cattle egrets ride on the backs of buffalo, elephants, rhinoceros, and hippopotamuses, getting a vantage point from which to see grasshoppers and other edible ground-creeping creatures. They are tall, keen-eyed birds, and from their high positions they are the first to see approaching danger. The instant they fly off, the big animals raise their heads, warned by the birds that something is wrong.

East African oxpeckers, which are starlings, and West African piapiacs, which are members of the crow family, both prowl the bodies of animals in search of ticks, maggots, bloodsucking flies, and suppurating matter from open wounds. When the oxpeckers are cleaning the bodies of impala, waterbucks, and gazelles, these antelopes often turn their heads and hold their ears steady so that an oxpecker can reach inside and yank out some parasite. When danger threatens, the oxpecker immediately runs to the opposite side of its host's body before it flies away, thereby warning the animal from which direction the danger is approaching. The behavior of the birds is neither completely commensal nor quite parasitic, since both the oxpeckers and the piapiacs can get their food from other sources. But it is a supportive relationship, which benefits both creatures.

The antelopes have a somewhat similar arrangement with zebras, which have better eyesight both day and night and see hunters first. And zebras are helped by the keener nostrils of the antelopes which can smell danger first. Again, neither creature really needs the other and when the dry season envelops them, zebra and antelope tend to separate, therefore reducing competition for scarce grass.

The practice of cleaning up another animal is common, particularly in the marine world around the coral reefs of tropical waters. There, larger fish, including the voracious morays, often line up for a chance to drift close to small fish or shrimp willing to clean them. The large fish gape widely, spreading their gills to let the cleaners reach in and find hidden parasites. The cleaning creatures work energetically and may cause the big fish considerable discomfort as they rip and tear at the objects attached to it, yet the fish is so anxious to be clean that it takes care not to make any sudden moves or to endanger its benefactors. The cleaners work efficiently until the job is done, and the big fish wait until the small ones have moved back into the safety of the coral.

When scientists investigated this phenomenon, they found that the big fish returned to the same cleaning sites every few days for regular servicing. If the cleaning creatures were removed experimentally, most of the big fish went elsewhere; those that remained, however, became noticeably fouled with external parasites. Once the cleaners were returned to the territory, the behavior pattern was immediately reestablished, and the big fish became sleek and healthy once more.

For those creatures that are neither commensal nor parasitic and can manage by themselves, a long-term relationship may be formed if one animal becomes at least partially dependent on another. Ants, for instance, use their antennae to stroke aphids, or plant lice, which respond by extruding "honeydew" from two tubes projecting from their abdomens. The honeydew is a waste solution of sugar and water left over after the aphids have processed great quantities of plant sap to get amino acids with which to make protein. Some species of ants become so fond of this sweet food that they "domesticate" the aphids. They pull them from leaves or stems and carry them to underground chambers, where the roots of the plants are exposed. The ants hold the aphids against the roots, and soon the aphids begin sucking the plant juices. The aphids, which reproduce by virgin birth, may so increase their numbers that they kill the plants they are sucking. The ants, however, are independent and can get along without the sweet treat of aphid honeydew.

Mexican carpenter ants adopt the caterpillars of a rare butterfly which lives in the pine-covered Tuxtla Mountains in southern Mexico. The ants catch the caterpillars as they hatch and dig a small burrow for each one close to its plant food. Guard ants occupy the burrow area, which may be up to fifty feet away from the dead tree where the ants are housed. At sunset, more ants go to the caterpillars' burrows and help transport them to the scrub, where the caterpillars feed and secrete honeydew, which the ants eat. At the first touch of dawn, the procession starts back to the refuge burrows, which are covered with earth pellets. The ants remain on guard while the caterpillars pupate into butterflies, and dig passages for the insects to escape, thus ensuring future generations of caterpillars.

Gary Ross, an entomologist from Louisiana State University, removed these caterpillars from their underground shelters and placed them on their food bushes with no ants to protect them. The caterpillars were just as hungry during the day as they were by night, and honeydew dripped from their bodies, but birds and lizards found and ate every one before dark. The survival of this rare butterfly, therefore, is now dependent on the watchful attentions of the industrious carpenter ants.

In the sea, some of the butterfish secrete a thick coating of slippery mucus that covers their flattened bodies and allows them to brush against the tentacles of such poisonous jellyfish as the Portuguese man of war. They are safe to hover under the dangling tentacles and feed on any creatures caught in the stinging cells. One of these butterfish, the man-of-war fish, grows to a length of three inches under a jellyfish with a ten-inch float. The slightest scratch in its mucous covering deprives the fish of its immunity and dooms it.

Damselfish, which also secrete a coating, live with large, colonial sea anemones among the reefs of the South Pacific. The fish are only about two inches long but they are conspicuously striped, often having bright blue bars crossing a tan background. Their striking colors and the way they cavort among the soft, thick tentacles of the anemones have earned them the popular names of clown fish and decoy fish. They certainly attract other fish, which dart after them and are immediately caught by the anemones. The clown gets its share of the meal while the anemone is slowly ingesting its victim. Demorest Davenport, a zoologist, and his associate, Kenneth Norris, experimented with clown fish and their anemones in sea-water tanks at Marineland of the Pacific. They discovered that the anemone must have a clown fish to brush against its tentacles; otherwise, it contracts them and stops hunting. And the conspicuous clown fish cannot long survive without the protection of the tentacles. The two unlikely creatures have become true partners.

From this kind of partnership it is but a short evolutionary step to another sort of ingenious deception. Some animals resemble and behave like others that are dangerous or evil-tasting. The harmless drone fly, which lives in the northern temperature zone, sips nectar from flowers and buzzes among the honeybees at work. It so closely resembles a bee in size, shape, color, and sound that no creature fearful of bee stings will touch it. So the fly is eaten only by honeybee hunters, such as robber flies and bee-eating birds.

Mimicry has been recognized since 1861 as a separate phenomenon. During the middle of the nineteenth century, a naturalist, Henry Walter Bates, found himself surrounded by mimics as he trudged through the Amazon basin. There were flies, beetles, and true bugs that resembled bees or wasps or ants with stingers. Some day-flying moths looked like bumblebees. Butterflies favored by monkeys or insectivorous birds possessed wings similar to those of poisonous or distasteful butterflies. Bates found that the mimics profited from this resemblance because their hunters had learned by experience to leave their models alone. But the mimics must always be less numerous than the models; they both must be abroad in the same area at the same time of day and year; and they must behave in a similar manner so that hunters cannot distinguish between them.

A German naturalist, Fritz Müller, who followed Bates into Brazil, suspected that there must be strong evolutionary reasons for a bee with a stinger to look like any other insect with a stinger. He discovered that the common encircling stripes of black and yellow serve as a united warning to all hunters, and so does the fact that most bees and wasps walk, fly, and buzz in a similar way.

In the great network of life the links between animals often extend beyond them to include the plants that give them food and sometimes shelter. Each creature and each plant is adapted to fit into the life system of the other. Short-tongued flies and beetles work on the flat-topped clusters of Queen Anne's lace and yarrow, or on the compact blossoms of spiraea and sumac. Bees and wasps with longer mouth parts reach nectar in deeper flowers, while bumblebees, which have long tongues and are strong, can force their way into snapdragons or suck nectar from deep clover blossoms. Moths uncoil even longer tongues to penetrate phlox and get to the bottom of columbines. Hummingbirds work on deep blossoms and seem to prefer those that are red or pink, whereas night moths search for large white flaring petals. The animals reward plants for attractive displays, since dusty pollen gets transferred from one drinking spot to the next, fertilizing the plants and enabling them to develop seeds.

Each plant species waits its turn to attract visitors. Bernd Heinrich, a zoologist, found that in a Maine bog bumblebees began the hunting year with a pussy willow, then shifted in sequence to leatherleaf, swamp laurel, rhodora and andromeda bushes, Labrador tea, sheep laurel, and small cranberry before concentrating three months later on meadowsweet, as they waited for the goldenrod to come into full bloom.

The flower may be able to seduce an animal that has little interest in either nectar or pollen. The orchids do this to perfection. They lure the males of some insects because they secrete a substance that closely resembles the scent used by females to attract males. The flower may have a colorful, velvety surface against which the male insect rubs himself, as though he were courting the female or actually mating with her. Temporarily satisfied, he flies off with a load of pollen, only to be seduced again by another orchid.

Some tropical trees have formed partnerships with ants. The ants get shelter and rear their broods of young, the tree is protected because the ants swarm out to bite and sting any animal that attempts to invade the tree or even touches it. In the rain forests of South America, the common trumpet tree can soar fifty feet, its hollow trunk pierced by ant holes. The moment it is jostled, the ants rush out to defend it. A trumpet tree which has not attracted ants is usually a tree without leaves since so many creatures are fond of its foliage. The voracious leaf-cutter ants can strip a trumpet tree in a day or two if resident ants are not present to defend it.

The leaf-cutter ants are also called umbrella ants, or parasol ants, and they, too, have a plant partnership with a fungus which grows on underground compost heaps built by the ants from pieces of leaf cut in the rain forest. These decomposing masses produce countless tiny balls of edible fungus which is practically the only food the ants and their young eat. When virgin queens swarm, they go first to the fungus garden and stuff themselves with the living fungus. Once a new burrow is established, the mated female begins a new garden with her own fecal pellets and tends it until she has raised her first brood of workers which will start another garden.

The industry of the leaf-cutter ants shows how a partnership between plant and animal can become a powerful force. On their forays out into the forest to find leaves, they leave wide tracks, almost like miniature avenues. These four-inch-wide roads wind through the forest for hundreds of yards and are marked by daubs of scent on trees where leaves may be cut. The half-inch-long ants cling to the edges of the leaves they have cut, carrying them upright like sails. Sometimes, a smaller worker climbs on the back of a porter and rides as a guard against parasitic insects and mites which try to slip past them into the ant nest.

There is a deep-seated relationship between many animals and different species of fungus. Their spores cling to the bodies of bark beetles, and so travel from one infected American elm to another, spreading Dutch elm disease. A related beetle uses a fungus as food for its larvae.

Green sponges and green hydras take their colors from the green algal cells embedded in their bodies. Though they are animals, they benefit from algal partnerships since the green cells provide them with oxygen and remove carbon dioxide, replenish the respiratory gas in solution and get rid of waste. Equipped with these green plants, and enjoying the sunlight, the sponges and hydras are more tolerant of environmental stress than when they are deprived of their partners.

Some of the simplest free-living flatworms appear to have formed an association with green algae. These flatworms are half an inch long and lack an organized digestive cavity, although they do have mouths. Each animal must find the spherical cells of a particular green alga and engulf them. The algal cells continue to flourish and multiply inside the flatworm, but they never become too numerous because the worm digests the surplus as its main food.

The reef-forming corals, all tropical creatures, produce limestone platforms at specific depths where the sun can brighten their world each day. The ability of the coral animal to secrete lime is linked to the vigorous photosynthesis carried on by the minute algal cells embedded in the coral's tissues. This partnership, which has gone on for at least five hundred million years, is the oldest known example of cooperation among neighbors. The cohabitation of unlike species, first recognized by De Bary more than a century ago, has become more than a scientific curiosity. Natural selection has rewarded these relationships, which probably began by chance and are now perpetuated by inherited patterns. In such complex and often incredible partnerships scientists can begin to perceive new dimensions to the phenomenon of inherited and learned behavior. And in such relationships we all see a common truth that is also a transcendental fact: One life is important to other lives.

The Deadly Hunters

Photographs by:
Alan Blank/National Audubon Society, 307
Edward R. Degginger, 319, 320
Harry Ellis, 318
Jean-Louis Frund, 321-323
Jessie Gibbs, 333-337
Clem Haagner/Bruce Coleman, Inc., 311
Grant Heilman, 308
John Hoke, 327
W.T. Miller, 309
Norman Myers/Bruce Coleman, Inc.,
328-332
Oxford Scientific Films, 312-314
A. Root/Okapia, 324-326
Edward S. Ross, 315-317
F. Sauer/ZEFA, 310
Caulion Singletary, 338

Methods of hunting exemplify complex levels of animal behavior. The kill is performed as an inherited response drive, automatic in its configuration and amazingly ingenious in its execution.

The lion, with its great strength, does not have to use all its power to make a kill. Instead, it has developed a rather clumsy but effective method of killing by suffocation. Many of the animals it hunts are too big to kill with sheer strength in any case. The lion knocks the animal down, then grips the stricken creature's nostrils between its jaws. A similar method is used by the constrictor reptiles. They do not actually crush their prey to death, as popular belief has it, but squeeze them until they can no longer breathe and they suffocate.

More ingenious methods are used by simpler forms of life. Some small fish shoot down flies with jets of water spurted from their mouths. The angler fish simulates a piece of food at the end of a body appendage which looks like a fishing pole. Some turtles attract curious fish by protruding their tongues, shaped roughly like a small fish, and then snapping up their victims. The kill has little, or nothing, to do with any human value system. Cannibalism is used if it is useful to the survival of the species. Unwanted or malformed youngsters are commonly eaten by their mothers, particularly in the cat family, and some kinds of crickets also eat each other.

Hyenas sometimes are eaten by their own kind during the great rush of the pack to get at the prey's body. The kill of the hyena is a study in efficiency. Everything is eaten. Tremendously powerful jaws smash up large bones and the broken fragments are swallowed whole.

Many animals coordinate physical skills with hunting instincts to capture prey. A hungry tree frog seeking a juicy fly, just out of reach, may leap and flick out its sticky tongue simultaneously (307). An African chameleon shoots out its tongue by use of hydraulic force, then captures the insect on the adhesive tip; the limp tongue and the prey are then slowly retracted to the chameleon's mouth (309). The edible European frog rests with its head just out of water, flipping out its tongue and capturing prey, such as a dragonfly, here half eaten (310). The sea star of coastal North Atlantic applies tremendous pulling power to force open bivalved mollusks. This blue mussel's shell is bent and held open while the sea star's stomach slides into the opening and ejects gastric juices into the victim to prepare for digestion to take place inside the victim's shell (308).

307

309

310

308

312

311

313

314

South African camel crickets, finding a dead member of their species, eat the remains (311). The tropical ogre-faced spider, *Dinopis*, throws its web at an insect (312). The victim is then jerked toward the spider (313), swathed in silk (314), and sucked dry. The tarantula hawk of the American Southwest, a large wasp, frontally attacks a large spider (315). The spider tries to avoid its enemy but the wasp rushes forward (316) and drags the body of its victim to a nest burrow (317). The wasp will lay an egg on the paralyzed spider, later to serve as food for the wasp's larva. The fishing spider of the American Southeast envelops her body in a film of air and enters a pond or stream to capture a small fish (318). The thread-waisted wasp, *chlorion*, captures crickets to stock the underground nursery, where its maggot-like young will feed on the bodies of the crickets until they are mature (319-320).

315

316

317

318

319

320

321

322

323

After an eastern garter snake has captured its wood frog prey, many slow movements are necessary for it to maneuver the food into its mouth. Most snakes have a lower jaw that temporarily unhinges so that they can work much larger animals into their mouths and gullets. If they have strong digestive juices which kill their victims (321-323), many kinds of snakes swallow their prey alive. The large South African tree snake, the boomslang, raids the nest of colonial weaverbirds, entering nests and eating eggs and young (324-326). At the edge of the Coppename River in Surinam, South America, an anaconda coils around a small caiman, a relative of the alligator, and suffocates it (327). The anaconda can reach a length of twenty-five feet and can eat animals of much greater girth than itself. Despite the great bulk of some victims, the anaconda's digestive system breaks it down slowly to liquid.

324

325

326

327

187

328

329

330

The hunt of the hyena combines the efforts of an entire pack of animals, each creature loping along at a tireless run to eventually overtake the exhausted victim. Then the kill becomes a communal effort, with each hyena struggling to get at the most desirable parts of the prey's body. The stricken animal, a wildebeest, one of the large African antelopes, is not killed directly but is literally eaten alive. The hyenas waste nothing, using their extremely powerful jaws to smash up the large bones, tearing the hide and eating it, and reducing the animal to skull and horns within minutes (328-332). The digestion of the hyenas is as powerful as their jaws, and the bones are completely broken down and expelled as a white, almost powdery mass. Sometimes the competition to eat is so strong that hyenas themselves are bitten to death and they, too, may be eaten in the mass carnage that follows the communal kill.

Until quite recently, the hyena was thought to be only a scavenger. But as knowledge of the creatures has grown, and scientists have perfected methods of following their nocturnal hunting, their roles as killers have been revealed. They work in clans, or tribes—usually dominated by a female—loose-knit groups of animals which increase and decrease in numbers depending on the time of the year. The best time of the year for their hunting is when the grazing animals—gazelles, zebra, wildebeest, and others—gather together to migrate in search of succulent short grass on the plains, brought up by heavy seasonal rains. But the females, with cubs in their dens, are restricted in how far they can hunt in a night. When the wildebeest are moving far and fast in pursuit of new grass, the hyenas may have to make long and exhausting runs in pursuit of them, or switch to scavenging.

331

332

189

Hunting for Fish

The egret of the American Southeast hunts and eats fish. It stands in a stream or lake, darts out when it sights a moving fish, and tosses its victim, headfirst, into its gullet so that the fish's scales do not catch in its throat. The egret's slender neck visibly expands to allow the fish to pass. The bird preens and rearranges its feathers before further hunting (333-337). The Florida alligator is almost omnivorous, and the large and powerful creatures have been known to attack deer unwise enough to be walking through the waterland prairies of the alligator's habitat. They also attack young turtles, raccoons, and any other small mammal which ventures into the shallow, weed-infested waters. They have even been known to dash out of the water and snatch pet dogs. But their favorite food is garfish, themselves rapacious underwater hunters (338), which are caught in underwater chases.

333

334

335

336

337

338

To seek food is a function of all living things. Both the hungriest lion and the simplest plant must secure food by whatever means they can. Within this searching framework there is an incalculable range of possible actions, linked to that distant moment more than a billion years ago when a simple sea organism slightly changed its behavior: Instead of absorbing organic materials into its body from the surrounding water, or capturing energy from the sun to form carbohydrates from carbon dioxide and water—the process known as photosynthesis—this ancient organism ate some other living thing. At first, the organism ate only plants, because the seas teemed with plant life, but the hunting success of this new entity was so great that its numbers spread around the world and launched a new way of life. The first plant-eating animals began hunting other plant-eaters. One animal ate another's carcass—or exploited it by robbing its food, by hunting it, or by feeding upon its living body as a parasite.

Animals live by these rules today. But, oddly enough, the great meat-eaters usually kill only when they have no easier way to get meat; they prefer to scavenge dead bodies, if these are available. This is particularly true of lions. The distinction between meat-eaters and plant-eaters, however, is blurred, because most carnivores eat leaves and fruit when the hunting is poor, and many herbivores eat insects, snails, and other small animals. Reindeer and caribou eat lemmings when these far-northern rodents go on one of their population rampages. Animals capable of cooperating with one another benefit from hunting in packs. They are thus able to kill and eat creatures that are bigger and stronger than any single individual in the hunting group. But the young of these hunting partners must be taught how to charge and to kill properly; the lessons often begin when small creatures that have been crippled by the parent animals are brought, maimed but alive, to the young as practice prey. The youngsters learn the techniques of the hunt by playing with the cripples before killing them. When they are skilled enough, they go on to larger quarry. Young wolves learn to hunt by running with their elders on caribou chases, but a mature wolf makes the kill. The youngsters do not touch the caribou until it is dead.

Rarely does a hunter attack unless it is likely to make a quick kill without much danger to itself. On the great African plains, hyenas, wild dogs, cheetahs, and lions are always on watch for any sign of abnormality among the hunted animals. A limping antelope will almost always be attacked, and the straggler is invariably cut down. The hunters have developed beautifully refined techniques to identify suitable victims. Cheetahs spend hours intently watching a nervous group of antelope, scanning every animal for clues that indicate weakness, and when the cats have made their choice, the attack is aimed specifically at a single vulnerable animal. The cheetahs will pursue it until it either is caught or escapes. Prides of lions and packs of wolves decide communally on a victim and frequently "test" it to discover whether it can outdistance them in a short chase. African wild dogs wear down their victims relentlessly by running after them until any weakness, whether of age or infirmity, is exposed. Solitary animals can be just as deadly as those that hunt in packs. The African leopard is a ferocious loner, surprising its prey by the swiftness of its silent attack. Its golden coat, spotted with brown rosettes, blends perfectly into the long grasses, or hides its presence in the foliage of a sheltering tree. When available, baboon is its favorite meal. Leopards are limited in their hunting territories since they prefer thick cover to open ground, and are programmed to hoard food. They frequently hide the uneaten remains of an animal in the forked branch of a tree, where the rotting carcass is sometimes found by vultures. Many other creatures, including man, share this hoarding instinct with the leopard, and store food for future use.

But the hunted creatures are not entirely helpless; they have a chance to bluff their enemies. The antelope always turn to face the cheetahs before they make their run, and sometimes the stratagem works; the cheetahs abandon their mission. The antelope closest to the cheetahs is rarely chosen as the victim. Hunted zebra bunch together in family groups, the stallions trailing behind and lashing out at their attackers with strong hind hoofs. Sick animals seek shade and shelter, especially to avoid the keen eyes of vultures, which can spot a stricken creature a great distance away and will wait days for it to die.

Smaller and more defenseless animals try to protect themselves from their hunters in a variety of ways. Some attempt to look bigger and more dangerous than they really are, while others pretend to be dead. One kind of shrew defends itself by swelling its back, squealing, and grinding its teeth. If these measures fail, the squeaking shrew falls backward, with its feet fluttering aimlessly in the air. The "death" defense is practiced with the most consummate skill by the opossum, which rolls over and "plays dead" frequently when it is chased. It is not certain what mechanism is at work when the opossum employs this strategy; the animal may be pretending, or it may be so frightened by the attack that its heart rate is slowed to the point that it faints. The opossum is not alone in feigning death to try to avoid the real thing; jackals, honey badgers, and striped hyenas are also able to look dead if necessary.

Millions of years of fooling the hunter's eye have demonstrated the benefits of being able to blend into the background, like the tropical katydids that have wings resembling green leaves. Keen-eyed hunters pick off victims that contrast with the background much more frequently than they do those that conform to it in color and pattern. Creatures chancing to inherit such beneficial adaptive characteristics are more likely to survive and to reproduce than those born without this capacity. The young of the survivors refine the change and perhaps extend it. For example, effective camouflage involves the creature's ability to remain absolutely motionless, or, as is the case with some marsh birds, to move their camouflaged bodies in harmony with a gently moving background of reeds touched by light winds. Many camouflaged lizards lie still as sticks until the very last moment before bolting from danger at high speed.

In the tropics, where bright and contrasting colors abound, the art of camouflage is highly developed. Multi-hued creatures or those with black and white patches in irregular patterns resemble blotches of bright sunlight and deep shadow. The moment an animal stops moving, its black parts join with its shadow to break up the recognizable shape of its body. A crow-sized toucan, with feathers the colors of sunlight and shade, alights on the branch of a tree and disappears from view. Small wonder that toucans must posture so elaborately to attract the attention of a female.

Many animals not protected by camouflage have other successful defenses. Since they do not blend into the landscape for safety, they move freely in full view of the rest of the world. Sometimes they are almost ostentatious as they display their markings, as if they were reminding hunters not to molest them.

The defense may be as simple as that of the milkweed caterpillar, which contains poison transferred from its food, or as complicated as that of the io moth caterpillar, which not only creates its own deterrent substance, but secretes it in the stiff, branching spines projecting from its body. The skunk's overdeveloped anal glands, combined with powerful muscles and nozzles, allow it to fire its stinking liquid at least ten feet. The bombardier beetle also fires a defensive liquid which it stores at the rear of its abdomen in two chambers separated by a closed valve. When the bombardier is attacked, it opens the valve so that the three chemicals housed in one compartment mix with the contents of the second. Quite literally, it is an explosive combination, and the liquid is noisily expelled from the beetle's body in a misty spray. The bombardier can aim its rear section in any direction and is able to fire repeatedly in its own defense. The hunter quickly turns away, but if it has been hit by the spray it is likely to suffer intermittent seizures afterward. Some marine animals and at least one flatworm that feed on stinging jellyfish and sea anemones are able to salvage the nettling cells intact, and then install these weapons on their own body surfaces.

Since relatively few birds store food, they are among the most voracious of hunters, their need for nourishment influenced by the fact that their body temperatures are from two to fourteen degrees warmer than those of other mammals. The insect-eaters, many of them songbirds, are especially greedy—more than four hundred plant lice have been found in the stomach of one chickadee, and a scarlet tanager has been observed eating about six hundred gypsy moth caterpillars in less than eighteen minutes. While many birds eat what is available, and change their diet with the season, some are specialized hunters, like a kite that inhabits the Florida Everglades and feeds only on one particular kind of fresh-water snail.

Some ducks swallow whole clams, pulverizing the shells in their gizzards. An oyster catcher waits until the shell opens, then uses its long pointed beak to cut the hinge muscles and get at the food. A Philippine kingfisher holds a snail in its beak and hammers the shell on a stone to crack it. One South American hawk sticks its beak into an open snail shell and shakes the creature out of its portable home. Some predatory birds lack the holding talons of an eagle, and so, like the East African shrike, they impale their prey on thorns before pulling them to bits. Several species hunt underwater for fish and are expert swimmers. Some pelicans scoop fish into their baggy pouches as they swim. Frigate birds take fish swimming near the surface and leaping from the water. Some long-necked herons stand silently in shallow water waiting for frogs or fish to come near enough to be speared, while other herons splash noisily through the water to disturb the resting fish. They eat the fish head first to prevent scales and fins from catching in their throats. Some birds resort to piracy to get food. Jaegers, which belong to the same sea-bird family as skuas, chase gulls and terns across the water to rob them of fish. Pursuing a tern, a jaeger will menace the bird with its hooked beak. The intimidated tern will usually drop the fish, or disgorge it, whereupon the jaeger will deftly catch its booty in midair.

While the fight for life and food is ferocious on land and in the air, it is particularly savage in the sea. Most fish prey on other marine animals, especially smaller fish and the young. Darting squid chase fish. Seals eat both fish and squid. Killer whales eat the seals. The great white pointer shark can swallow a man whole. A thirty-six-footer caught near Sydney, Australia, had teeth two inches long. Younger white pointers and smaller sharks bite their victims into several pieces so that they can swallow the bits. Sharks are also scavengers. They help to rid the seas and estuaries of carcasses that smaller fish cannot devour so quickly. They come close to shore at night and swim into the turbid waters of outflowing rivers searching for carcasses on their way to the sea. The largest sharks are the least dangerous, rarely troubling man unless they are molested. Two species reach more than forty feet in length: the basking shark, which filters out drifting animals from plankton, and the whale shark, which takes small fish and crustaceans swimming in large schools. These great sharks often weigh four tons. Their success as food-gatherers is based not on speed or strength, but on their persistence in exploiting all the available food as they laze along through marine waters.

Sharks are an anachronism in the modern biological world, their ancestry going back almost unchanged for one hundred million years. Their primitive skeletons are still made of cartilage. They cannot make themselves buoyant without swimming. The moment they stop, they sink gradually to the bottom, and when they are at rest, water no longer flows through their mouths and out between the gills to aerate their blood. A motionless shark is soon a dead shark. This is why, in large marine aquariums, a diver must "walk" a newly captured shark until the anesthetic wears off and the creature is able to swim on its own. Like the sharks, the sea snakes are reminders of the very distant past, surviving largely unchanged in form and habit. Many of the fifty-odd species known today are so deadly poisonous that skin-divers who encounter one have no protection against its attack. Death occurs within seconds of being bitten.

Crocodiles and their near kin, the alligators and caimans, preceded the hunting sea snakes in evolutionary time, but these creatures, echoes of the ancient past, remind us that they are descendants of the successful hunters of long ago. The crocodiles have had to combat countless modern problems, including human hunters, and in most parts of the world they have fared much less well than the marine animals. All crocodilians dig or build nests to conceal their eggs, and many of them remain near the nest to guard against its many enemies. The Nile crocodile is vulnerable to the predations of the monitor lizard, which has a special skill for locating and digging up crocodile eggs.

Little was known about the behavior of these animals during their early years until the eminent English zoologist Hugh B. Cott checked the decline of crocodiles from Uganda to Rhodesia a few years ago. Cott discovered that young crocodiles rarely expose themselves in open water where they will almost surely be eaten by larger crocodiles or other enemies. Instead, they race through the shallows like lizards, exploring weedy areas and small pools, climbing bushes, and hunting for water beetles, small crabs, insects, and spiders.

As a crocodile grows, it seeks larger prey in the tangles of papyrus stems and vegetation at the shoreline. It catches big crabs, frogs, and toads, and crushes snail shells before swallowing them as extra nourishment. Occasionally a young crocodile manages to catch a small bird, a fish, or a rodent drinking at the edge of the water. As it becomes better able to defend itself, it risks expeditions into deeper water and becomes more nocturnal. Although the crocodile still basks by day, closing the vertical pupils of its eyes to slits, it develops into a night hunter with wide-open pupils and with every sense alert. Through its adolescence, the crocodile begins to chase catfish and other fish prowling the river bottoms near the shore in search of food for themselves. But the crocodile is not an enemy of the commercially valuable fish; rather, it continues to hunt the hunters of these fish and thus helps to preserve the African fishing industry.

At about ten years of age, the crocodile matures into the seasoned hunter that will eventually dominate any stretch of water it occupies. Now about five feet long, it becomes sluggish and does not eat as frequently. Instead, it attacks larger victims, striking them down when they come to the water to drink at night. It can reach up and haul down a large python hanging from a shoreline tree. It pursues its constant predator, the monitor lizard. It waits with only its eyes showing above water for antelope and zebra to come and drink from the still water of its pool. Then it fires itself forward like a torpedo, skidding up the bank next to the startled victim. The crocodile, with a lash of its powerful tail, hurls its prey into the water. There the crocodile subdues a large victim by rapidly rotating its body with the animal held between its jaws. The crocodile has no trouble holding its breath; it clamps shut its valvelike nostrils while its victim drowns. The tremendous force of the twisting action usually tears the animal in half, although even a big crocodile may have difficulty ripping up a full-grown zebra. But once the victim is dead, there is no hurry. The body will decompose, and the crocodile does not care whether its meat is fresh or putrid.

In its maturity the crocodile is almost completely wide-ranging in its taste. It eats young buffalo and hippopotamuses, soft-shell turtles and fish-eating birds, cobras, and geese. Children who wander into its waters are taken immediately, and so are domestic livestock. In some areas, stockades have been built so that women can do their laundry and children can play in safety.

The Nile crocodile and the American crocodile are both at home in the sea. The African species has colonized the island of Madagascar, nearly two hundred and fifty miles from the mainland, though it has never become a true sea creature. The real saltwater crocodile, which swims between the tropical coasts of Asia through the East Indies to northern Australia, is a much bigger and more powerful creature. It can grow to twenty feet, compared with the African animal's twelve-foot length. Both the American crocodile, which ranges from southern Florida and the West Indies into Central America and Colombia, and the Orinoco crocodile grow up to twenty-three feet, the longest of any of these creatures. They represent a hereditary line of predators that goes back 175,000,000 years to when they were contemporaries of the earliest dinosaurs.

The jaws of a crocodile, which are capable of being opened so wide and are so strong that they clamp an almost unbreakable grip on any victim, link it with practically all the other meat-eaters. A lion's jaws, gripping the spine of a buffalo or wildebeest, clamp together and crush bone, cartilage, and flesh, though the lion often kills its victim by fastening its teeth around the animal's nose and mouth and suffocating it.

The killer whale's mouthful of teeth and the power to drive its cylindrical body forward at high speed give it an unmatched ability to kill at sea. The teeth and powerful jaws of the dragon lizard of Komodo Island in the East Indies give it a similar power to subdue victims on land. This largest of the monitor lizards grows to ten feet and weighs three hundred and fifty pounds. The hard, sharp edges of the parrot fish's mouth produce a cutting force to bite coral animals out of their limy cups on a reef. A snapping turtle's bite can drag down a duck.

The mouth parts of the praying mantis, the caterpillar-hunter beetle, the ladybug, and the dragonfly are no less powerful, relatively speaking, though their sharp jaws come together from two sides, instead of from above and below. Even the immature dragonfly displays the power of these biting jaws. Much of the forward part of its head is hidden behind its lower lip, which is doubly hinged and can be extended far out in front of the young creature to seize a tadpole or a small fish. This peculiar arrangement allows the young dragonfly to conserve its energy, since it waits for a victim to come within range before it reaches forward and strikes.

A frog or a toad catches an insect more quickly than the eye can follow by flicking out a sticky tongue and hitting a bug, which adheres to the tongue. The tongue is whipped back into the hunter's mouth and the bug is tossed down the short gullet into the stomach. When a frog or toad tries to eat a larger victim, such as an earthworm or a big caterpillar, it uses its forelegs to keep the squirming creature lined up lengthwise so it can be swallowed.

The African chameleon can extend its cylindrical tongue, which wraps its tip around the victim and holds it securely. Then the tongue goes limp, and the chameleon appears to swallow it. The chameleon's accurate tongue-throw is guided by extraordinary eyes, which operate independently and watch the world from both sides of the body.

Chameleons and toads all display the parallel forms that evolution takes when a good hunting technique is developed. There is no conceivable relationship between chameleon and anteater, but both use long sticky tongues to hunt. The giant anteater of tropical America has a tongue two feet long; the unrelated African aardvark has a one-foot tongue. These creatures use their strong claws to rip open a termite nest or a mound of ants, then slither their wet tongue in among the confused insects, reaching far down into branching galleries to collect food on the sticky surface. To operate such extraordinary organs, these creatures need exceptional muscles. The scaly anteater's tongue muscles run back through its body to its hips, giving it the strength to force the tongue back and forth at blinding speed. In one stroke, the clean tongue is covered with ants.or termites.

The most primitive of the snakes, which also happen to include the largest, have their own specialty. The pythons and the boas kill their prey by coiling around their victims so tightly they suffocate. The largest member of the boa family is the anaconda, which lives in northern South America and grows to about twenty feet. The reticulated python, which inhabits the wet tropic lands from Burma to the Philippines, weighs about 250 pounds and grows to twenty-eight feet. These snakes detect their victims by extending their tongues to capture samples of air and carry them to the sensitive pits in the roofs of their mouths.

Snakes' reptilian ancestors had feeble jaws. Forced to swallow their victims whole, these early creatures possessed shoulder girdles, which prevented them from swallowing victims that were too large. Now all snakes disengage the hinge joints of their jawbones so their mouths can stretch wide enough to accept their victims. The snake slips one side of its mouth, and then the other, around the body, gradually working the meal down its throat. Snakes no longer possess shoulder girdles and front legs, and among most species one lung has been suppressed to give more space for the passage of a victim. Sometimes when an African python swallows a sharp-horned gazelle, the horns pierce the reptile's body, but the python is affected only if they make a serious rip in its outer skin.

These hunter snakes apparently evolved quickly at a time millions of years ago when there was a simultaneous rapid evolution of rodents on which the snakes preyed. Both hunter and hunted flourished because the seed plants were greatly encouraging the development of mammals at that time. Every group of great hunters must have evolved during such ideal conditions to reach the delicate balance they have achieved with the animals they hunt: neither destroying all the creatures on which they depend nor starving to death themselves.

At least one fish developed its ingenious hunting method after the appearance of insects on earth. This is the seven-inch archer fish of Asia and Australia, which swims in shallow water and waits for an insect to land on an overhanging leaf not more than a foot above it. The moment the insect is suitably placed, the fish rises and fires a jet of water at it. If the fish misses but does not alarm the insect, it fires again. When the insect falls into the water, it is snapped up and eaten. Such a hunting skill must have developed over long periods of trial and error.

An equally long period of evolution must have been spent developing the hunting flatfish. A hatchling flatfish, such as flounder, halibut, plaice, or turbot, is only a few days old when it ceases to swim normally. It can no longer keep its belly down and its back up. One side of its head begins to enlarge faster than the other. One of its eyes gradually migrates across the top of its head toward a position close to the other eye. At the same time, the young fish begins to lean as it swims. Eventually, when its eye has finished the migration, the fish settles on the sea floor on the side that has no eye. Its mouth has assumed a peculiar twist and its coloration has changed. The eyeless side resting on the bottom remains white, but the top side, with the two eyes, takes on mottlings and hues that blend it into the appearance of the sea bottom. The fringes of its flat body hug the bottom so snugly that no shadow appears, no matter how shallow the water. This makes the flatfish a deadly hunter.

The ungainly goosefish attracts its victims by using its "fishing pole," the modified spine of the dorsal fin. The pole carries a flaglike flap of tissue at its tip. The goosefish wags this lure back and forth above its mouth until it gets the interest of a victim. With the victim suitably placed, the goosefish abruptly opens its mouth to create a sudden inrush of water, and the victim is sucked through its sharp teeth, which are then snapped shut. These extraordinary hunters, which grow to be four feet long and weigh about forty-five pounds, can be found from New England to Brazil. About one hundred and twenty kinds of angler fish swimming in the dark ocean depths use similar techniques. Victims are scarce so far down, but the poles of the deep-sea angler fish have luminous lures; sometimes the lining of the angler fish's mouth is also luminous. Fish, squid, and crustaceans are all attracted by the strange lethal lights.

But perhaps the most ferocious and complicated hunting techniques are those used by predatory insects. The bulldog ants of Australia can bite with one end and sting with the other. The army ants of tropical America and the driver ants of Africa are capable of moving a hundred feet an hour, overwhelming and cutting into tiny pieces almost every living thing in their path. The larva of the predatory tiger beetle, which lives in arid, sandy places, tunnels through the hot sand at an angle. The hungry larva waits at the top of its shaft until an insect comes within range. Even a large insect cannot pull the tiger beetle larva from its burrow because attached to its abdomen are two hooks that are embedded in the walls of the tunnel.

Assassin bugs are spread around the world, and each species has its own hunting technique. The members of one species cover their legs with sticky resin from coniferous trees and then raise the legs to trap insects. One West Indies assassin exudes a substance that is attractive to ants, but when the ants drink it they become disoriented and helpless.

But the best defense against being eaten lies in the strict rules of the hunt itself. The "victims" are not actually pursued very often because the hunters are programmed to concentrate on that small minority that are easiest to kill—the young, the old, and the sick. Despite their predators, grasshoppers teem in the dried trash of summer; millions of caterpillars let themselves down to earth on silken strings; heedless birds fly overhead; immense herds of antelope and horses feed on the African plains; enormous schools of fish travel together to breeding places. When mankind does not tamper with the delicate balance between prey and predator, the laws of natural selection operate as a restraining influence on the hunting creatures, and as an evolutionary goad for the hunted.

The Challenge of the Future

Scientists who study fossils calculate that about one per cent of all the kinds of life that have lived on earth and in the seas during the last five hundred million years are still alive. As members of this surviving elite, and as the only species that has learned to compare past and present, we must consider ways to ensure our survival on this crowded planet. No other kind of life has entered into partnership with so many different plants and animals or so thoroughly exploited the resources of the world. We used to believe that the ingenuity of our ancestors freed us from the restraints imposed upon other forms of life. But we know now that this was a false presumption and that we must somehow restore the system we seem to have lost.

Perhaps we can return to our proper place in this diverse world if we come to understand our role in it. Perhaps if we watch how other forms of life manage their own survival, we will be able to apply these patterns of behavior to the human situation.

The dramatic episodes in animal behavior—the fights, the runs, the kills—are much rarer than most people generally believe. Scientists who have recorded the total time spent by individual animals in such activities have come to the conclusion that nonhuman life *usually* avoids conflict. Even meat-eaters kill only when hunger triggers action, and then the victims are usually taken from the minority that are sick or old, or from among the surplus young that are eating too great a share of the limited food supplies. Plant-eaters, too, generally do not eat all the leaves and flowers on a plant that sustains them. Animals frequently draw back from their environment and from other creatures, conserving energy for a future time when action would be more profitable.

The populations of creatures flow irresistibly into posterity but in this dynamic torrent of activity, births and deaths need not alone control the size of a population. Surplus creatures may emigrate and actually reduce the total. If the death rate is particularly high, the population may be boosted by the immigration of creatures that were surplus elsewhere. The amount of attractive space that is available to each animal governs the size and distribution of the population, and the density of creatures, in turn, has a powerful effect on births and deaths, emigrations and immigrations.

To study the dynamics of populations, it is necessary to see them in all phases of their growth, stability, and decline. To find a population that is both scattered and well-fed, we must look for colonists moving into new territory. We may find them where the tundra has just been warmed by the sun after a long winter; where a flood plain lies deep in silt left by a river that has recently crested to its highest level in a century; or on an abandoned farm. In these places, seedlings and sporelings burgeon in an abundance of moisture, sunlight, soil, and nutrients. The plant-eaters find plenty of good food, and their hunters may be slow to catch up with them. There is not likely to be trouble from parasites or disease as long as the nourishment is so plentiful. The creatures eat heartily, grow rapidly, mature early, and begin reproducing. They may bear more than one new generation during this benign season, and an unusually large proportion of the offspring will survive and repeat the process. But space and the growing season are both limited, and these factors usually control the growing animal population before its demands exceed the supplies of food; thus are the plants saved from overexploitation.

When the animal population has passed its peak, its numbers remain fairly steady. Death and emigration balance birth and immigration. The creatures eat less, grow more slowly, mature later, and are less successful at reproducing. This situation exists even when the supply of food is still ample.

African elephants, for instance, adjust their birth rate in response to some hidden cue from the environment. In equatorial regions the highest annual birth rate—80 to 90 births for each 1,000 adult animals—is in areas where there are only about two animals per square mile, on the average. The lowest birth rate—between 60 and 65 per 1,000 adults—occurs where a few more than four elephants live in each square mile. A research team from Washington State University has discovered that female elephants become sexually receptive at ten or eleven years of age when they live in uncrowded areas. Conversely, where the population is dense, the cows may not mature sexually until they are twenty years old. The intervals between pregnancies also differed in the two areas; the intervals were longer in the crowded environment.

Since bull elephants rarely fight over females and maintain no permanent bond with them, the population-control mechanism seems to be an inborn feature of the cow's behavior. When the conditions are not right, she stays an adolescent longer and remains close to her mother, avoiding mature bulls. When she finally becomes a mother, she nurses her youngster longer, even though this means she must ignore her own sexual cycle that recurs every six months. This is the closest thing to birth control that any animal has evolved and one so effective that it does much to ensure the continued welfare of the elephant populations. Their survival, though, is threatened by their enormous appetites and by their inability to find enough new feeding grounds in modern Africa. Each adult elephant eats about a ton of selected shrubs, tree foliage, and plants every day, and when elephants are living together in a large herd, they often eat faster than the plants can grow. Shade trees and other woody plants are quickly destroyed. Fire follows their destruction, and grass-cropping antelope move in. What was once lightly treed savannah country becomes open grassland. For the elephants, this seems a permanent change, and they emigrate, which often allows the vegetation to recover—within a few decades. In primeval times the system worked well enough, but today African populations are rising so rapidly, and so many farms are spreading over the continent, that the elephants have very little new land left to exploit. Behind them lie thousands of square miles of devastated territory. No one who has ever seen land destroyed by elephants is likely to forget it. Shrubs wrenched from the soil lie with their roots exposed while the dead wood of smashed trees dries in the hot sun.

The elephants have no mechanism to allow them to adjust to conditions imposed by an alien influence, and they go berserk when they cannot find new food supplies. The plight of these huge creatures reveals not only the complexity of animal societies but also the pervasive influence of man upon them. At Kruger National Park in South Africa the elephants must be controlled if the park is to survive. The elephants were no problem when the land was set aside for a park in 1905. About ten elephants lived in the region then, the survivors of decades of tusk-hunting. But with their safety assured inside the park, the elephants multiplied to 500 by 1945, and numbered 1,000 in 1960. Since only certain areas of the park are suitable for elephants, park managers estimated that the limit was about 2,500 elephants—one for each three square miles of park. At that time, as many as 50 elephants were entering the park each year across the poorly fenced boundary separating Kruger from Mozambique and southern Rhodesia. Park officials built a strong fence to reduce this immigration and began to shoot about 100 old elephants and sickly young ones each year to compensate for the normal increase from new births. But by 1964, officials found there were nearly 2,400 elephants in the park, and in 1968 there were more than 6,000. By October, 1970, the figure had risen to more than 8,000 animals.

In Kenya, a couple of thousand miles to the north, the elephants got caught in a similar dilemma, but this one was brought about by drought, as well as man. The drought forced them to make extensive migrations, and hundreds of them died of thirst and starvation. The survivors destroyed all the trees in some park areas and caused major environmental changes. Before the settlement of East Africa, elephants made tremendous migrations, some of them getting as far north as Ethiopia. But today wandering elephants encounter hunters, angry farmers, townspeople, highways, and railroad lines almost everywhere they try to wander. Their long evolutionary history prepared them for population control, but not for the advent of man.

Crowding affects other animals as well as elephants. In the American tropics, the six-foot-long iguana lizards assemble along a river to mate and lay their eggs. Normally, these reptiles live high in the rain forest feeding on foliage and basking in the sun, but when the time comes for reproduction, each female must find one of the scarce areas of sunlit sand along a river where she can dig a nest, drop her eggs and cover them, and let the warmth of the sun incubate them for her.

Space is limited, and soon every suitable site has been used at least once. Late-arriving females simply dig out the buried eggs to lay their own, and vultures eat the eggs kicked out of the sand. In the confusion of overlapping digging, laying, and covering, many future iguanas die, but there is also a fairly consistent hatching of survivors. The critical factor is how much sunlit sand is available to the females.

In the Antarctic, Weddell seals face an equally critical moment during the coldest part of the winter. The freeze is so solid that the animals have difficulty keeping their breathing holes in the ice open. Eventually they must take turns opening each hole to get a gulp of air. Their pups are born between October and mid-December, and by the end of March the young seals must compete with larger and stronger animals for a place at the breathing holes. This hidden conflict beneath the ice keeps the population constant. The females can give birth to any number of pups but the environment will only accept a limited number. The survivors stay fat and healthy because there is always enough food to go around.

The rush for housing between adults and young is fierce among the Northern Hemisphere marmots, woodchucks, and ground hogs. For the first year, the young stay with their mother in her burrow, occasionally playing on the bare earth at the entrance. In late summer they follow her on foraging expeditions. By fall every surviving youngster is as fat as its mother and able to withstand the fast of hibernation until spring. When warm weather arrives, their mother evicts them, accepts a temporary mate, and soon gives birth to another litter. Now the young animals must find havens for themselves, and quickly, in a world where every well-drained slope, every sheltered burrow may already have an occupant. By fall all available shelters are filled, and those animals left without homes will die. The limits of housing possibilities, therefore, control the population, and there is no need for mature animals to fight with one another.

Herring gulls do not interact so peacefully on their nesting grounds. Each pair of parents seems to be constantly threatening the nearest neighbors with beaks and voices and wings upraised in menacing gestures. But there, too, territory is vital. The gulls occupying the center of the gullery are the older and more successful birds; younger birds are forced to take territory around the periphery of the gullery, where attacks by enemies are much more common and defense of the nest much more difficult. Although they defend the gullery en masse, herring gulls frequently eat eggs and nestlings of their own kind if they get the chance. The population controls of the herring gulls have been distorted by man, as were those of the elephants. The herring gull has always been a circumpolar scavenger, its success limited only by the available carrion. But as human communities have increased in size, the herring gulls have become bolder; our garbage dumps, sewage systems, and fish-processing plants give them a bounty of surplus food which they are supremely able to exploit. This has caused a major population explosion and forced the herring gulls into many new environments. Some have taken to pirating blueberry crops or raiding vegetable gardens. Other thousands attack colonies of petrels, kittiwakes, and terns, killing their fellow sea birds for food. Despite determined herring-gull control programs, the bird remains buoyantly tied to the conditions of its new environment, which are so far beneficial, and it will continue to prosper until its numbers become too great for the environment to support.

Among the higher animals, stimuli from a crowded environment are supposed to work through the central nervous system to affect reproductive rates. Sex hormones may be involved in a secondary role. But among the lower animals it is possible that chemical control is more direct. The waste products from a large population may affect the reproductive mechanism. J. B. Best, the distinguished professor of biophysics at Colorado State University, noticed some years ago that free-living flatworms showed a surprising awareness of their environment. In solitary confinement, they displayed a strange "lethargy" that resembled boredom. Best learned to cure the ennui by offering the worm extra space. He found that a population of worms tended to stabilize at about 600 to 900 creatures in a pan of water. The worms got all they could eat and had their water and pans changed regularly, yet they cut down drastically on their breeding, despite what appeared to be ideal conditions.

Best and his co-workers took ninety worms from the crowded culture pan and put thirty of them into a four-inch bowl of water. Almost none of them reproduced during the next ten days. The second group of thirty worms had the rear third of their flat bodies amputated and were also put into a four-inch bowl of water. They recovered from the surgery, but almost none of them divided. The last batch of thirty worms lost the front third of their bodies, the part containing eyes and the simple nerve center that served as their brain. Within seven days, fifteen of these worms divided, and more followed suit in the next ten days. All the worms were regenerating their missing parts, although none of them yet had eyes or a new nerve center. Prohibition against dividing in crowded conditions had something to do with the front end of the worm. As long as the mechanism remained intact, the animal could hold reproduction in dense populations to a minimum.

Quite another kind of control is demonstrated by many of the grazing animals, and it may be characterized as one of quality rather than of quantity. The buck impala's rounding up of a large harem of does, and his subsequent frantic defense of them against all the other bucks without harems, is part of a system that confines breeding to certain selected creatures. This may have a direct bearing on controlling the populations of these fecund animals.

Many other African grazing animals also reward dominant males with the fruits of mating. The unused adult males have been called "unemployed breeders" by V.C. Wynne-Edwards, the regius professor of natural history at Marischal College of the University of Aberdeen and a scientist who is well known for his studies of Scotland's red deer. The exclusivity system involves confrontations that establish, and then reinforce, the relationships between dominant breeders and those that must remain bachelors.

Such selectivity is fundamental to almost all animal societies, but very little is presently known about how it works. The many animals that use scent to mark territory give us some hint of the system's complexity. Scent signals are vital to night animals, which cannot easily communicate by visible or audible means. Many rodents use urine, secretions from special glands, and even smears of vaginal fluid to mark territory. European rabbits and some larger animals such as the black rhinoceros defecate in carefully selected spots to inform later arrivals of the marker's proximity, sex, and perhaps even its physical condition.

Some African gazelles use a special gland located just beneath their eyes to mark territory. The gazelle thrusts its head down onto a dry stalk of grass, which enters the gland hole and is smeared with the black, tarry marking substance. All the cats urinate at the limits of territory, and hyenas defecate to mark their passage across hunting territories. For the hunting animal with sensitive nostrils, the dark night is probably a veritable encyclopedia of information about creatures in the area.

D. D. Thiessen, of the psychology department at the University of Texas, has studied rodents from northeastern China that use special sebaceous glands hidden in the fur on the undersides of their bodies. Both sexes use this gland differently. A female Mongolian gerbil smears her young with a secretion from the gland, and as she hunts for food she lowers her body to mark the ground, something she does most frequently when she has a nest and young to nurse. Her sex hormones regulate this marking behavior as they do her aggressive defense of the nest area. Male gerbils must work harder under the stimulus of their sex hormones. The male hurries around a new territory and marks it three times as often as most females. During these marking trips he will fight any other male he meets. If he wins the fight, he retains the territory. If he loses, he cedes the territory to the other animal, becomes submissive, and goes into hiding. He will not breed as long as he can smell the scent of the victorious animals, but if he can reach fresh territory, he resumes marking immediately and is ready to fight once more. Memory still haunts him, though, for he stops marking immediately if he catches one whiff of the scent of the animal that beat him. Thus, his entire role as a breeding animal can be turned on and off through his nose.

The smell from marked territory reaches our nostrils in old houses where mice and their odors have been long confined. The smells of mice are full of messages and have a profound effect on the behavior of entire populations. One of the strangest of these responses is known as the Bruce effect, after the British zoologist H. M. Bruce, and it appears in certain strains of house mice. The moment a pregnant female smells the marking of a strange male close by, she reabsorbs her embryos and ends her pregnancy. Apparently, she learns the odor of her mate and remembers it. Strangely, the effect of the odor of the alien male is less three hours after mating than it is when it is introduced a day later. Somehow, the memory of her mate must be given time to mature.

The unknown substance that produces the Bruce effect is a component of male mouse urine. A pregnant female shows almost no reaction if a drop of her mate's urine is painted on her nose, but a drop from a strange male, even if it is diluted with distilled water, causes the Bruce effect. Young female mice quickly mature sexually if they only smell male adults, or if their noses are painted daily with male urine. Conversely, they are sexually retarded if they live in an all-female group, and their sexual development is slowed if urine from female mice is painted on their noses.

The whole structure of animal activity may depend on a constant reception of the right signals. And when the signals are not received, or are transmitted in a faulty fashion, disaster may strike. House mice sometimes go on breeding sprees in the fields and rise to overwhelming numbers. This has happened several times in California, alarming both householders and farmers. Every time a plague has occurred, cats were imported and pest-control programs instituted, but neither of these measures has ever had much effect. Instead, the mice controlled their own numbers. Just as fast as the populations rose, they crashed, until only a few mice survived. This phenomenon seems connected with the famous lemming plagues in Scandinavia and the far north generally, but both remain unexplained. The only common connector is the fact that at the peak of population and during the crash practically none of the females are pregnant.

Lemmings, which resemble meadow mice with short noses, legs, and tails, have survived in the harsh realm of Arctic tundra and barrens. Most of their terrain was covered with glaciers during the last ice age and through those long cold periods the ancestral lemmings must have traveled widely and frequently. Their survival probably depended on finding small patches of suitable territory, either where the advancing ice had killed back a forest or where retreating ice had left some exposed land on which evergreen trees had not yet begun to grow. The sharp contrasts of climate probably kept the lemmings' numbers low, but when conditions improved slightly, the animals had to be able to boost their populations rapidly and thus take advantage of the brief good times. Their fecundity also gave them a chance to create large emigrant groups that could go out in search of new land to colonize, thus ensuring that the lemmings were spread over the largest possible area. If this explanation is correct, the population sprees of these tiny creatures were more important several thousand years ago than they are in the more settled conditions prevalent today. An Australian mammalogist, J. J. Christian, feels that the ancestral territory is very important to the modern animal. History, he suspects, determines whether present behavior will check reproduction or cause frequent population explosions. Ancestors that lived in large forests or grasslands, or in the sea, rarely needed to cross inhospitable country to reach territory similar to their own. These animals probably adjusted their numbers to remain slightly below the maximum that their habitat could support. Among their descendants Christian looks for social hierarchies and signs of stress from overcrowding, such as the enlargement of the adrenal glands or spleen. But animals whose ancestors lived in unstable times and constantly exploited new territory would be better served by periodic population explosions, rather than routine control, Christian feels.

He suspects that the meadow mouse is prolific because originally it was an opportunist that quickly took advantage of a changed environment. Perhaps it moved into the meadows that were formed when abandoned beaver dams collapsed. The beaver ponds drained, and their silty bottoms were quickly colonized by meadow grasses. Very soon, shrubs and trees would invade the region, shading out the grasses, and the meadow mouse would be driven off. During the lush years, the mouse colonists needed to build up large populations and disperse their surplus members in all directions. If only a few of these mice found another new beaver meadow, they would be enough to save the species from local extinction.

The reproductive capabilities of these tiny creatures are one of the underpinnings of all life where meadow mice live. Each meadow mouse eats about twenty-three pounds of hay a year and can produce up to eight youngsters in a litter, which are ready to breed only eight weeks after birth. Between March and late fall, a female meadow mouse may produce a dozen litters. If all her youngsters lived and reproduced themselves without mortality, she would be responsible for more than 235,000 descendants before winter came. But there is an external control: Countless hunters suppress these enormous numbers of mice.

Despite the killing, the mice often outbreed their hunters and crowd up to 12,000 of their kind into one acre—that is one mouse for each four square feet. Such population explosions devastate their meadow world and force the mice to emigrate.

Christian has described another kind of colonist that plays a significant role in spreading out its species. These creatures, the unemployed breeders bested in competitions with dominant males, sometimes wander beyond familiar territory to quite distant areas. If they find a haven, they will recover their sexual drive and be able to act as true colonists. They have a chance to throw off the traditions of the old territory and launch new ones.

The lemmings' great numbers encourage the survival of northern hunters such as owls and foxes. But when the lemming population crash comes, these hunters are desperately hungry. The Arctic foxes wander more widely and show increased daring in their hunting tactics. The snowy owls, which thrive on lemmings at any time of year, can migrate south, which they usually do once every five years or so. In North America, they sometimes come as far south as New Jersey, and they can reach Portugal on the European continent. They hunt meadow mice and voles on these southern migrations, but their arrivals are unpredictable and no one knows whether they return to the Arctic in the spring or perish in the south.

Whatever their fate, the owls are part of a massive distortion begun by the eruption and subsequent crash of the lemmings. In the spring, large flocks of snow geese and blue geese wing northward up the Mississippi Valley from wintering grounds around the Gulf of Mexico. They head for their nesting places beside tundra pools. There they meet the Arctic foxes that have survived a winter without lemmings. Fox and owl collide in mutual hunger, but the snowy owls are too alert, too strong, too sharp of claw and beak, and too ready to attack any fox for the mammal to take advantage of the bird. Instead, small colonies of snow geese, black brant, and eider ducks cluster their nests close to the regular perches of the owls, and are able to rear their broods where foxes dare not intrude.

In the forest sanctuary of Isle Royale in Lake Superior, quite a different community of animals interweave their lives on their 210-square-mile home. Moose are the main victims of the timber wolves, with beaver as a supplementary diet. The wolf population numbers between twenty-two and twenty-eight animals at midwinter, but this increases slightly when the moose have a good year and produce more calves, because the young wolves have a better chance to reach maturity. The moose population is fairly stable at around 600, which is just the right number to browse the available vegetation without damaging it.

The island can hold about 140 beaver colonies, each with a pair of parents and three or four young. Every time mothers give birth to new litters of kits, the two-year-old beavers are driven from the family. Because the island is so fully occupied, few of these banished animals find homes, and the wolves get most of them. Those beavers that manage to crowd into locations along the limited number of flowing streams are forced to cut aspen trees farther and farther from the water, and so expose themselves increasingly to the hunting wolves. The surviving beavers come to depend more on aquatic vegetation and other foods less nutritious than aspen bark, thus giving the aspens a chance to regenerate. Every five years or so, tularemia, a virulent fever transmitted by ticks and fleas from the island population of snowshoe hares, strikes the beavers. This gives the aspens another chance to sprout from old roots, grow into seedlings, and build up new stores of food for future generations of beavers.

Meanwhile, the population of hares may fluctuate sharply without affecting the basic mixture of larger animals, though lynxes and foxes may suffer setbacks when the hare population is down. The island community is essentially closed, forcing all its inhabitants to interweave their lives in many ways and providing us with an example of balanced populations. Largely because of man's interference, and the complexity of most systems of life, such balance is rarely as stable as it is on Isle Royale. Intensive studies of animals in their native territories, surrounded by familiar neighbors and places, have convinced scientists that neither predators nor parasites alone arbitrarily control populations. Even outright starvation is not especially important. Instead, the controlling mechanism is usually set by social behavior. Other causes may include the "birth control" techniques of females, bad weather striking creatures at the limit of their range, and the onset of periodic plagues.

Scientists now realize the importance of the trigger that releases populations from their previous behavioral patterns. When one group of animals successfully use a new kind of response to a crisis, they survive in ways that are different from the rest of the population.

When the house mouse reached New Zealand by ship in the nineteenth century, it found a mouseless paradise. Although cats, weasels, ferrets, and rats were also introduced to New Zealand, the mice spread rapidly. Some populations established themselves in houses, following the English house mouse tradition, while others split away to begin new lives in tussock land, in farm pastures, in wheat and barley fields, in forests of introduced American pines, and in dense native hardwood forests. These separate groups already are beginning to show differences in behavior that are moving them further away from their kin. Soon the differences will be great enough for them to be split into races or subspecies, and New Zealand may one day have half a dozen different species of mice, all descended from the one type that settled the country originally. Each population will have its own small world to conquer, its separate welfare to maintain.

Isolation and opportunity create a marvel of evolution, fitting new forms of life into new or changed places. Coexistence can only be bought at the expense of suitably changed behavior. Neighbors change and adjustments are needed. The old ways are not good enough for the new situations of today and tomorrow, nor is any one way the best way. Conditions never remain uniform from place to place, or from time to time, and all simple answers eventually become inadequate or downright contradictory. There is no constancy in the grand design of life.

Man cannot predict the future, and mice can only hint at what happened in the past. The smallest and largest forms of life must live with their separate heritages and accept the consequences of every choice. But in diversity lies our best chance of enduring. Each day brings a new opportunity to modify and balance life for more satisfactory forms of survival.

Index

Numbers in italic refer to captions and illustrations

213